MATERIALS SCIENCE—
SELECTION OF MATERIALS

MATERIALS SCIENCE–
SELECTION OF MATERIALS

S. W. JONES
Assoc. Eng. (Sheff.), C.Eng., M.I. Mech. E.

LONDON
BUTTERWORTHS

THE BUTTERWORTH GROUP

ENGLAND
Butterworth & Co (Publishers) Ltd
London: 88 Kingsway, WC2B 6AB

AUSTRALIA
Butterworth & Co (Australia) Ltd
Sydney: 20 Loftus Street
Melbourne: 343 Little Collins Street
Brisbane: 240 Queen Street

CANADA
Butterworth & Co (Canada) Ltd
Toronto: 14 Curity Avenue, 374

NEW ZEALAND
Butterworth & Co (New Zealand) Ltd
Wellington: 49/51 Ballance Street
Auckland: 35 High Street

SOUTH AFRICA
Butterworth & Co (South Africa) (Pty) Ltd
Durban: 33/35 Beach Grove

First published 1970
© S. W. Jones, 1970

ISBN 0 408 70110 2
ISBN 0 408 70111 0

Filmset by V. Siviter Smith & Co Ltd, Birmingham
Printed in England by The Whitefriars Press Ltd,
London and Tonbridge

CONTENTS

PREFACE

A particular feature of the many changes in the constitution of first degree courses in engineering over the last decade has been the emergence of subjects which attempt to combine the elements of physics, chemistry and metallurgy with the practicalities of manufacturing processes. Contact with students at many levels suggests that difficulty is found in these Materials Science subjects in striking the essential balance between classical concepts and manufacturing feasibility.

The object of this volume is to provide a text illustrating how available physical data and knowledge of production methods can be combined at a sufficiently early stage in the design process so as to make a significant contribution towards optimum selection. It may be read as a companion volume to 'Engineering Design Problems', (Iliffe, 1969) and while discussion is retained on an academic plane, the requirements of the engineer whose formal training did not include assessment of material properties have not been overlooked. The data provided will be of assistance in preparation for the descriptive sections of the C.E.I. examination in production technology and contains many references to topics found in first year syllabuses for other examinations involving consideration of production processes.

It is impossible in a book of this size to provide a comprehensive text on the full range of forming processes and the procedure adopted is to indicate trends rather than process data, so that diagrams are drawn to indicate variations in dependent parameters rather than specific values. The tenor of the discourse is in terms of statement, argument and decision, typifying the design attuned attitude which the author is trying to instil.

The early chapters are concerned with an explanation of the basic strength and property problem for metals and how forming methods, with the help of subsequent treatments, can be chosen to satisfy a particular specification. A review of non metals precedes the final chapters which are specifically orientated to bearing and gear materials. In order to provide a satisfactory coverage for these transmission components, it is necessary to explain how design fundamentals will influence material and process selection and alternative design methods have been discussed in some detail.

In order to cover the transition of changeover from Imperial to metric units, the SI conversions are shown for numerical values where appropriate. This practice is not recommended for general student use since the philosophy underlying SI thinking and its preferred number values ought to be assimilated as soon as possible. The conversion values are therefore provided, without rounding off, as a temporary expedient to assist the mature reader during the period when familiar parameters and sources of data are being presented in unfamiliar guise.

A selected bibliography is included to stimulate further reading together with a series of test questions representing varying degrees of difficulty. These have been adapted from many sources, including past examination papers, and assume knowledge of related disciplines in a first degree course. In some cases further study of the recommended reading matter will be necessary before the solution can be found.

S.W.J.

1
MATERIAL PROPERTIES–DECISION THEORY

In this introductory chapter the student is shown the application of a systematic working plan for materials selection. The fundamental design method is the application of intellectual and creative ability which, when allied to correct choice of material, results in proposals for a product, not only to meet the specification, but also to permit manufacture by the most economic method.

Such studies involve consideration of material properties, how such properties will affect the manufacturing process and the implementation of a scheme of evaluation which is a function of any design method.

1.1. TYPICAL ASSUMPTIONS

The strength property affects the ease with which a material can be deformed to the required shape and specifies an ability to resist loads in service. The simplest design process is restricted to the use of test data for yield strength and relies on conventional assumptions such as:
1. Elasticity and homogeneity of material.
2. Absence of plastic deformation below the elastic limit.
3. Within the elastic range a uniform distribution of direct stress or a predictable distribution from other loading systems.

The designer utilises a Safety Factor to allow for variations from an ideal situation, to act as a safeguard against contingencies due to calculations based on incomplete data, or for the situation where a simple solution of the problem is not apparent.

In more realistic terms the design process ought to start earlier by considering the process method for the raw material and the effect of the process on design properties. Typically ductility must be considered, representing a deformation capability, as many metal

	Tensile strength	Hardness	Ductility
Grey C.I.	——	———	
Mild steel	--------------	------	-------------
Nickel	———--———--———	——--	———--———--—

Fig. 1.1

working processes are limited by the available ductility of the material. A useful pictorial guide to properties of common materials can be compiled similar to Fig. 1.1 which is drawn to scale.

1.2. USE OF TEST DATA

Close consideration of mechanical test results reveals a general ignorance of the performance of materials in practice. It is rarely possible to predict behaviour following the onset of yield, unless there is a careful control of load values and time scale in the plastic range. Without such data it is difficult to apply the mechanics of forming processes, although the majority of such processes produce a compressive stress situation. Thus, even if a specimen may fracture under a small extension in a tensile test, the same material can be forged to many times its original length under favourable conditions of heat and pressure.

A cautionary word needs to be expressed as to the danger of using test data when knowledge of the test conditions or failure hypothesis is incomplete. The most useful data is that which relates findings to test piece dimensions, rate of loading and environment and the translation of results into working parameters must take into account the differences between careful preparation of a test piece and the manufacturing method for a much larger component. The ruling section referred to in BS. 970 and 971 provides an indication as to whether desired properties will be available in a finished component.

The problems of utilising test data are well illustrated by considering material behaviour in a simple tensile test. The cross-sectional area decreases rapidly when plastic deformation begins (necking of specimen); the load required to continue deformation falls off but the true stress σ_T is always larger than the nominal stress σ based on original dimensions $A_0 L_0$. The relationship between these stresses is shown in Fig. 1.2 and it is often convenient to plot

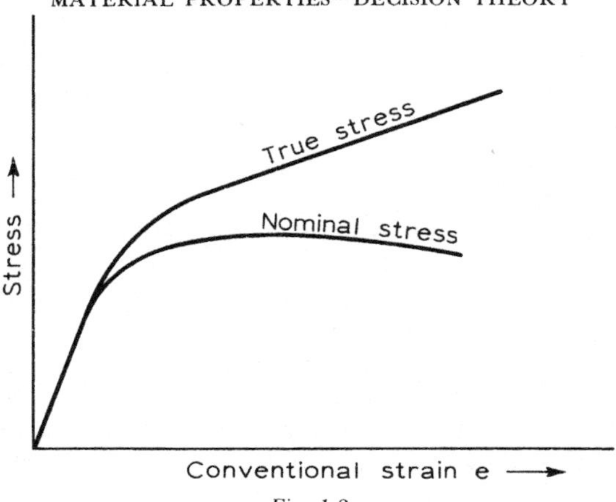

Fig. 1.2

a true stress–true strain curve or flow curve, so called because it emphasises the plastic range. True strain

$$\epsilon = \ln(L/L_0)$$

and the relationship with conventional strain e is given by

$$\epsilon = \ln(e+1)$$

Neglecting any volume change

$$A.L = A_0 L_0 \text{ so } \epsilon = \ln(A_0/A) \text{ or } A_0/A = e+1$$

True stress $\sigma_T = P/A = (P/A_0)(A_0/A) = (P/A_0)(e+1)$

Now σ_T is conventionally written as a function of e so

$$P = A_0 f(e)/(1+e)$$

by differentiating for dp/de and considering that maximum load occurs at a strain when dp/de is zero we get

$$d\sigma_T/de = \sigma_T/(1+e)$$

Fig. 1.3

and the construction shown in Fig. 1.3 is obtained. The tangent cuts curve at true stress value corresponding to maximum load and the σ_T ordinate at the maximum nominal stress (conventionally U.T.S.).

It is often convenient to regard the loading cycle as statistical in nature, since it can then be represented by a distribution curve. Comparison of curves for applied load and load capability will reveal any possibility of overlapping and any tendency towards a failure situation. In such terms, Safety Factor is defined as the ratio of mean values of load capability and applied load, related to some type of failure occurrence; yielding would be a typical postulation.

Life considerations must be applied to a real situation; so called designing for infinite life is rarely possible with a new product, all that can be done is to design for a failure rate which will be acceptable.

1.3. DECISION TAKING

In the planning of new designs there will be a number of ways to reach the objective once a basic shape has been formulated. The solving of decision equations as a computer programme will provide optimisation of parameters such as size, geometry, material, quantity. Nevertheless, the selection of material and process is a formidable

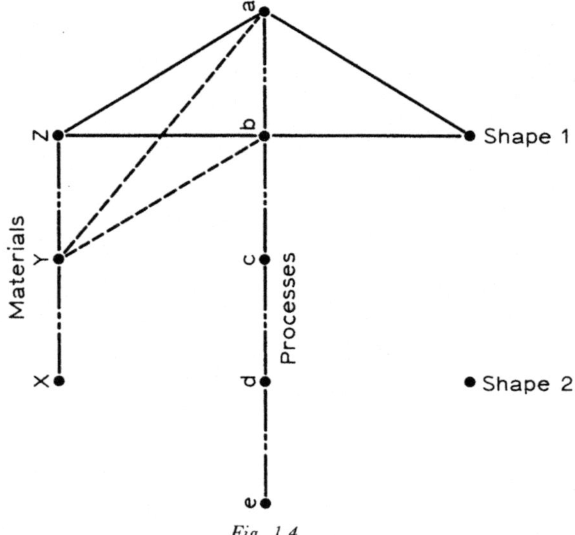

Fig. 1.4

task, taking into account a possible choice from 20 000 metal alloys.

Decision trees of the type shown partly completed in Fig. 1.4 are most useful provided that the dependence between alternatives is recognised. Such diagrams provide a good exercise in the implementation of decision theory.

Remember to allow for the probability associated with each possibility, a substantial reduction of the decision problem comes from recognition that the probability of a particular occurrence has an insignificant value.

A number of proprietary materials data systems are now available which list properties obtainable from particular alloys for various heat treatments and processing methods. The data is expressed in terms of ratios of properties such as weight, strength, cost, etc., so that an optimum choice is clearly discernable from the list of possible materials, once the desirable criteria have been established.

The young designer must not believe that a suitable solution will present itself by some mysterious or accidental means in due course. Solutions are stimulated by organised thinking around the problem, which might involve examination of previous developments or patent specifications; while the application of a methodical work plan will point the way, it is no substitute for intellectual ability.

There are a number of decision capabilities which can be taught, but the success rate of implanting the ability to logically deduce, or to exercise fine judgement in a choice between materials and methods is notoriously low.

EXERCISES

1. Illustrate how a product material specification can be satisfied by charting a score of mechanical and non-mechanical functions with relative rating factors.
2. Construct a design tree involving two materials, five processes and five shapes, calculate the number of design decisions involved and show how ranking will indicate the best strategy.
3. Review factors involved in the assessment of Value Engineering and show how it can be applied to commercial Product Design.
4. List three criteria which define the end of a product's effective life. How can curves of natural distribution be utilised to assess the likelihood of product failure?
5. Show how Safety Factor can be defined on a statistical basis and derive typical equations to estimate its value.
6. Explain why care must be taken in implementing the results of conventional test data and discuss the proposition that a torsion test may provide a more fundamental measure of plasticity than a tensile test.

2

MATERIAL STRUCTURE

It is not necessary for the designer to have a specialised knowledge of metallurgy, but the fundamentals which govern material properties demand an appreciation of the physical behaviour of atomic structures and of phenomena experienced in the melting and forming of alloys. With this background the basis of possible relationships between micro-structure and mechanical properties can be assessed, although it is dangerous to base working properties on purely empirical formulae derived from micro-structural behaviour. With one constituent, either pure metal or solid state solution of two metals, concern is with the size and shape of grains. With two constituents it is necessary to consider the properties of each phase in terms of volume and distribution.

2.1. BEHAVIOUR OF ATOMIC STRUCTURES

An atom is conventionally regarded as a positive core or nucleus around which are orbiting series or groups of negative electrons. The chemical properties of elements are related to the number of electrons in the outer group and, when this group is incompletely filled, the tendency to form stable groups results in a gain or loss. The placement of atoms can be regarded as points arranged periodically in three dimensions. A solid metallic state is recognised by a geometrical arrangement called a space lattice in which the repulsion due to positive charges at lattice positions is balanced by negative charge of electron cloud which holds them together. Alternatively, consider an arrangement of *ions* in a mean or average position about which there is thermal oscillation. Motion increases with temperature, ionic volume increases and the characteristic thermal expansion of a solid is experienced.

Some solid materials demonstrate allotropy, with a different lattice structure at different temperatures and each structure having different properties. Such phase changes demand diffusion which

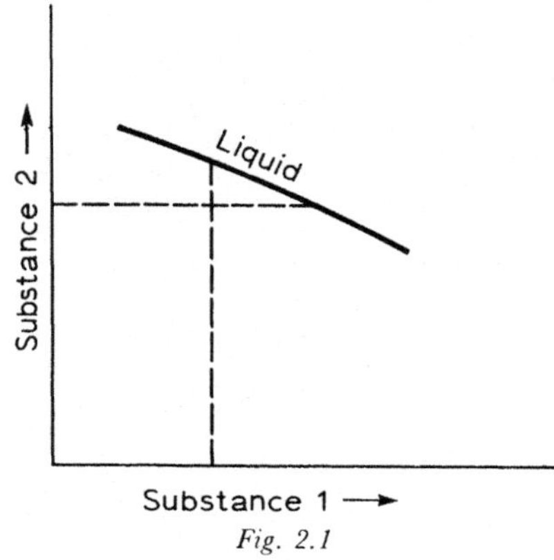

Fig. 2.1

occurs along boundaries much quicker than in grains. Fig. 2.1 represents two substances which are soluble in the liquid state; the freezing point of each is affected by adulteration with a horizontal line giving composition by proportion. At a eutectic point the liquid is saturated with both solutions and can precipitate crystals of each. When the metals are insoluble in solid state two kinds of crystal are formed each having their own cell arrangement. The lead/antimony system is a simple eutectic, as the constituents are soluble as liquids but not as solids. Crystallisation of most metals results in lattice arrangements called body or face centred cubic or close packed hexagonal. The two latter may be regarded as densely packed spheres whose co-ordination number (representing nearest neighbours) is 12, whereas the less closely packed b.c.c. number is 8.

2.2. METALLURGY OF FERROUS METALS

Fig. 2.2 represents the space lattice for iron and carbon known as face centred cubic; notice the identical surroundings and the closest possible packing. Notice also that the diameter of a space is smaller than a carbon atom so that local distortion is caused when such an atom tries to join the group; this prevents entry at any adjacent space.

Tendency of two metals to form a solid solution is determined by relative size of atoms and a similar lattice pattern. A knowledge of crystal behaviour is the key to performance of a metal in a forming process. Classification of type is in terms of bond between structure units and, since different directions are rarely equivalent geometrically, we need to take account of their anistropy with respect to

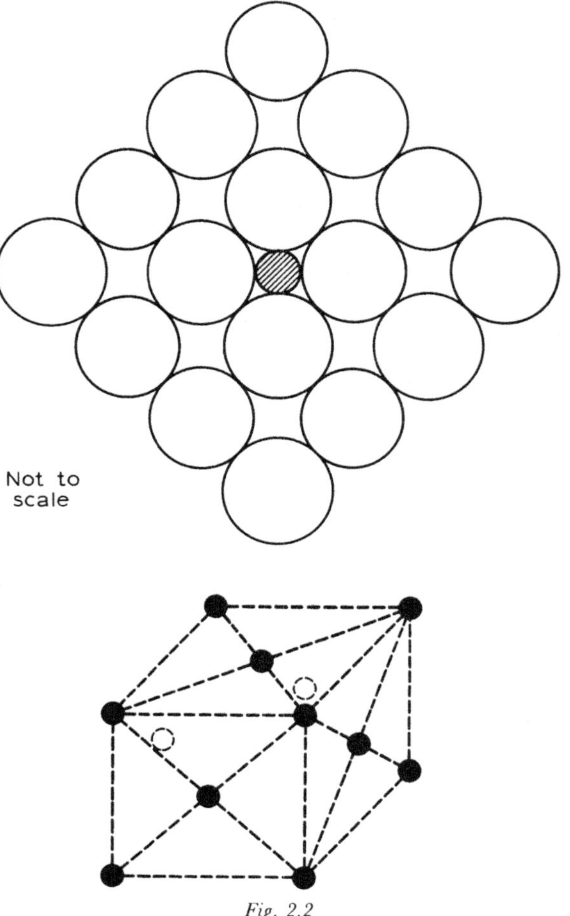

Fig. 2.2

physical properties. For example, cubic crystals are isotropic with respect to electrical resistivity but not in terms of elastic deformability. Each type of crystal has planes on which slip may occur and certain directions of sliding.

For the f.c.c. type referred to earlier there is one type of slip plane and twelve possible slip systems. The simplest assumption of slip envisages a plane sliding over its neighbour at some position. This concept demands a constant shear stress and it is not found in practice where one area slides before another, as a continuing cycle in some metals and slip recommences on another set of planes as they approach a certain position.

The boundary between slip and a whole region creates a physical situation of lattice imperfection called a dislocation. Since the strength of a single metal crystal is little more than one two-hundredth of the strength required in a commercial metal there is evident need for some type of strengthening process. Various methods are in use, obstruction of dislocations is very common, solid solution alloying, precipitation hardening and work-hardening techniques are other examples. The basic strength calculation is on the assumption that aggregate properties are the average of free crystal properties and total is a summation based on distribution. We speak of a metal having strength when slip is confined to a short run whereas a metal which does not slip is regarded as brittle.

Metallurgical considerations enter into the choice of systems providing resistance to wear and to the desirability of improving appearance by some type of surface finish. The many mechanisms of wear make it almost impossible to guard against every eventuality. Property variables, environmental state and degree of lubrication are typical factors. Manipulation of constituents or subsequent treatment can be used to provide a hard skin or a wear resistant surface and potential difference between dissimilar metals and other corrosion aspects should not be overlooked.

The two basic roles of a surface finish are to provide resistance to deterioration and to stimulate the sales appeal by colour and appearance. The designer must organise within the limits set down by customer specification, production requirements and selling price. Design considerations might involve a choice between organic, electroplated, anodised, galvanised or hard-faced finishes and apart from any aesthetic values, unusual characteristics such as reflectivity, heat resistance or conductivity may have to be taken into account.

The metallurgical behaviour of ferrous materials is simply illustrated by considering iron-carbon mixtures. Equilibrium phase diagrams might be regarded as a basis, although in practice their value is limited since equilibrium states based on slow cooling are rarely found in commercial steels.

A conventional cooling curve, as shown in Fig. 2.3, might lead to consideration of crystallisation as similar to freezing of water, but it is not applicable to pure molten metals as an unstable supercooled region will develop if there is no solid matter present. This results in the formation of a dendritic growth whose point of contact

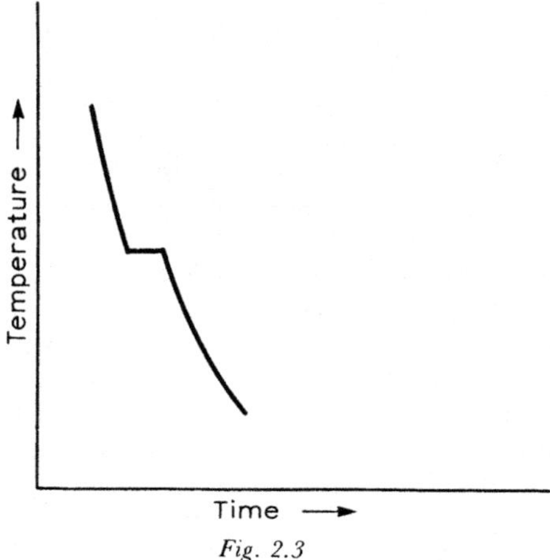

Fig. 2.3

with a neighbour becomes a crystal boundary.

An alternative structure representation, as shown in Fig. 2.4, is by means of the inverse rate curve, showing the change in heating or cooling rates at particular critical points. Changes here reflect an allotropic situation with the possibility of reversible changes in atomic structure and corresponding property changes. A well-known example of phase or state change is that of pure iron whose constitution, body centred at room temperature ($\propto$ iron) changes to face centred at about 908°C (γ iron) and returns to body centred

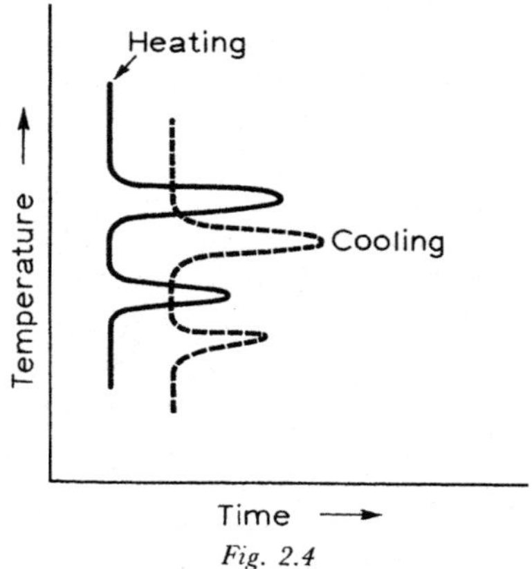

Fig. 2.4

(δ iron) at about 1403°C. Of more concern is the presence of carbon in solution, which not only changes the transition temperature but also the range of temperature over which phase changes take place. Fig. 2.5 represents part of the iron end of a typical equilibrium diagram; notice how the temperature for the start and completion

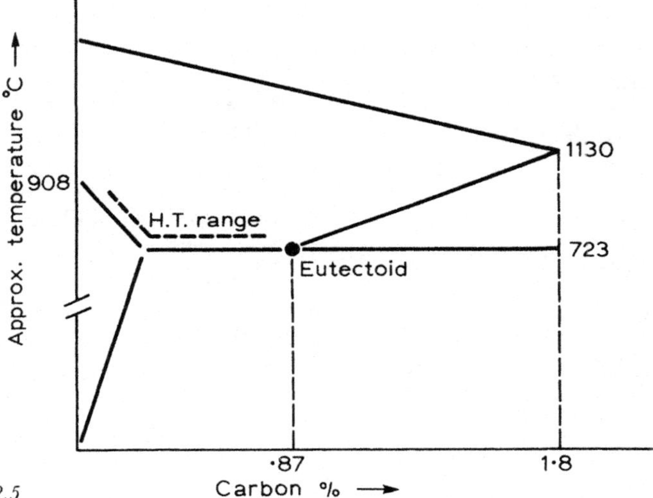

Fig. 2.5

of the change to f.c.c. is reduced as the carbon % is increased, also how separate solid phases go into solution to form a new phase, without a change in temperature. The increase of strength coupled with reduction in ductility is shown in Fig. 2.6 as a marked result of the addition of carbon. Perhaps the most important facility is the

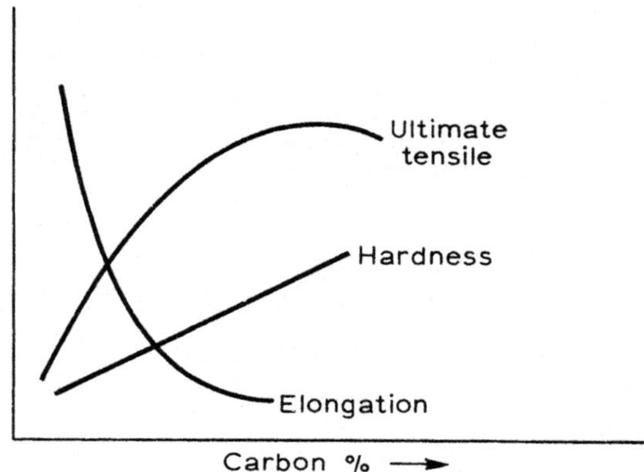

Fig. 2.6

availability which is offered for hardening, as the metal is rapidly cooled from an elevated temperature. Structure alteration by heat treatment is considered further in Chapter 8.

A plastic deformation means an irreversible change in the unit cell and it is necessary to distinguish between an idealised situation where each volume receives the same distortion and the configuration as in a tensile test where bar ends are gripped. Metal behaviour can be characterised in terms of the shear-hardening curve; in f.c.c. types shear hardening does not keep pace with increase in shear stress so necking commences. However, ultimate stress and elongation depend on the initial single slip which is the reason for anistropy of these properties.

In hexagonal crystals the shear-hardening curve has an end point, since fracture takes place after a certain amount of work has been done on the crystal and properties are therefore uniquely defined by orientation. The slope of shear-hardening curve is temperature dependent, at high temperatures recovery may be completed during deformation cycle but shortening the test time increases apparent shear hardening.

EXERCISES

1. Explain the basis of crystalline growth and how certain characteristics of a metal may be predicted from knowledge of its lattice structure.
2. Discuss the types of bond between metallic crystals and how these influence the mechanism of elastic and plastic deformation.
3. Classify common metals in terms of their lattice structure and explain the effect of dislocations.
4. What are the important differences between slip in a crystal and in a polycrystalline aggregate, how is the grain boundary effect reflected in performance?
5. Compare the effect of grain boundaries and slip lines in terms of initiating fatigue cracks. Discuss the effect of size and speed of loading on fatigue behaviour.
6. Explain the difference between the construction of a ceramic and a metal with particular reference to the type of bonding and resistance to slip.

3
SELECTION CRITERIA

In the rare cases where a single design consideration uniquely defines material selection there is no need for a logical process of choice. Attempts are sometimes made to relate such simple systems to applications involving cyclic variations of load, temperature and operating time by means of 'play safe' considerations of the theoretical worst case. This leads to pessimistic solutions and a typical case of over designing will result.

3.1. FUNCTION OF THE SELECTION PROCESS

Once tentative sizes are known, the selection of a material may pre-dispose the manufacturing method. Conversely a decision to use a cheap and convenient manufacturing method should involve a feedback to the process planner since it will influence the preparation of the raw material. The shape and dimensions of a casting in terms of a minimum section thickness is a typical example.

It is thus necessary to define an acceptable relationship between function, size and cost; a readily available weaker material which forms easily may be an acceptable solution provided that increased size and weight does not lead to spatial or support problems. A typical evaluation might be in terms of considering the cost of additional machining to provide a specific shape from a block of cheap material, compared with the cost of casting a more sophisticated material so that minimum subsequent machining is necessary.

A common mistake is to consider cost as the only criterion; load capacity in terms of size or weight is a useful parameter and recent developments in the field of lightweight alloys have provided a range of materials whose strength/weight ratios are superior to the more common ferrous materials. Logarithmic plots of price/lbf and price/in³ yield interesting results; using dimensional criteria or an equal volume basis the material cost of an aluminium alloy design may be 30% cheaper than mild steel, although 13% dearer in

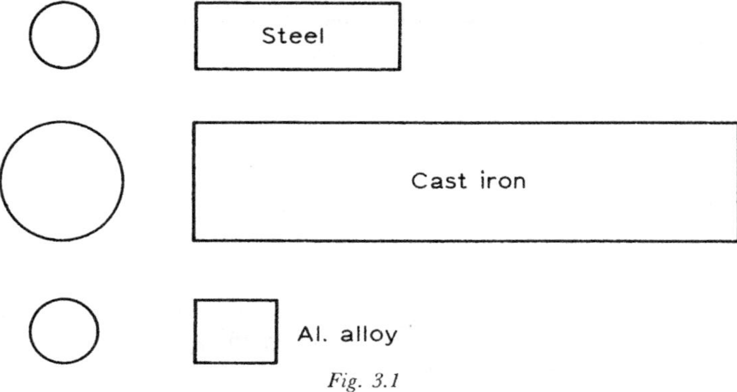

Fig. 3.1

terms of equivalent strength. The real design technique is of light-weight form rather than the application of lightweight materials to a conventionally designed shape.

The maximum utilisation principle leads to the idea of not providing material where it is not needed, adjacent to neutral plane in a rolled section is a typical example. Useful tables can be drawn to compare properties and the dimensions under various types of loading and Fig. 3.1 shows dimensional comparison for an identical

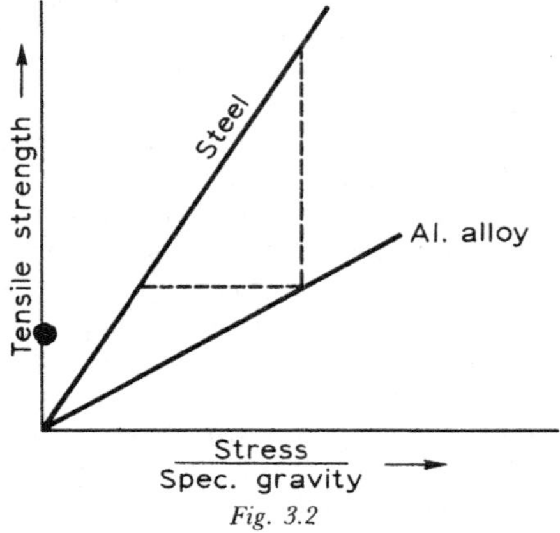

Fig. 3.2

tensile load on each material. Fig. 3.2 shows that for the same tensile strength the weight saved is in the ratio of specific gravity.

The situation may be reached where a high-grade alloy steel has the same structure weight as in light alloy but we approach then the design region where full utilisation of high strength steels theoretically permits such thin sections that their stiffness is open to question. The use of lighter materials however implies that particular attention be paid to stiffness, with safeguards against buckling and unwelcome deflection due either to load or thermal expansion. Further consideration of light alloys will be found in Chapter 9.

3.2. ACCURACY OF MANUFACTURING PROCESSES

In any discussion as to the provision of a required quality or maintenance of a desired reliability, consideration must be given to manufacturing tolerances, interchangeability and methods of inspection. It would not be appropriate to include in this book a detailed analysis of limits and fits or the mathematical background contained in assessment of the probability of inaccurate manufacture. If, however, insufficient attention is paid to the basic inaccuracy of manufacture and the natural tendency to vary from desired limits of size shape, form and finish, much work in the choice of material and forming process will be wasted.

3.3. CONTROL OF PRODUCTION PROCESSES

The basis of statistical control of a production process demands that assignable causes of error are removed so that any variations are due only to random causes. Some control methods are based on the premise that the frequency of error occurrence in terms of its magnitude follows a normal distribution which can be defined in terms of a standard deviation. In quantity production, the extremes of a size range occur infrequently and it may be more economical to increase the tolerance range and accept the possibility that a small percentage may lie outside prescribed limits of size.

Naturally, if tolerance limits are set narrower than the natural spread the manufacture of defective parts is inevitable; one empirical basis is to set the tolerance range one-third greater than natural spread.

A cost/tolerance relationship similar to Fig. 3.3 is a useful guide to ensure the best selection from the choice of manufacturing processes available. Quite apart from any restrictions imposed by the method of measurement, the satisfactory allocation of tolerances demands an estimate of clearance extremes which will be satisfactory for a particular application and some knowledge of the accuracy obtainable from manufacturing plant.

BS. 1916 forms the basis of the standard method for determination

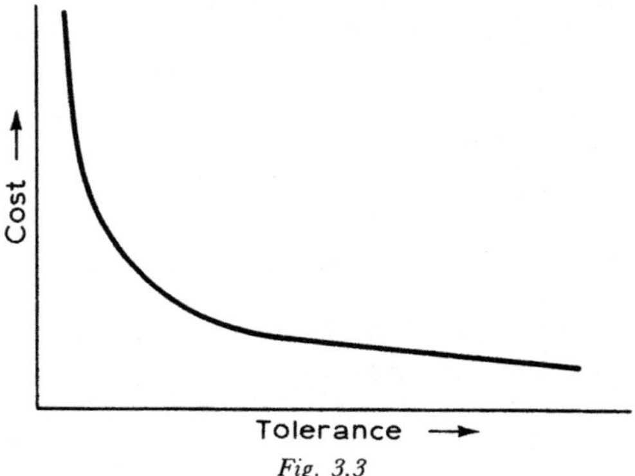

Fig. 3.3

of size tolerance and requires correct utilisation of the fundamental tolerance unit (FTU) which in 0·001 in units$=0·052D^1/^3+0·001$ (D) where D is the geometric mean of the diameter step involved (inches). Alternatively FTU in microns when D is mm$=0·45D^1/^3+$ $0·001$ (D). Some scheme is needed for increase of tolerance with size and grade of quality specified, which is an indication of the degree of difficulty in manufacture. The size tolerance for each feature is obtained from multiplication of FTU by a preferred number range where each step is approximately 60% greater than its predecessor. In order to achieve the desired clearance the variation from design size is calculated from a fundamental deviation.

Fig. 3.4 shows typical relationships for a clearance fit. Unless

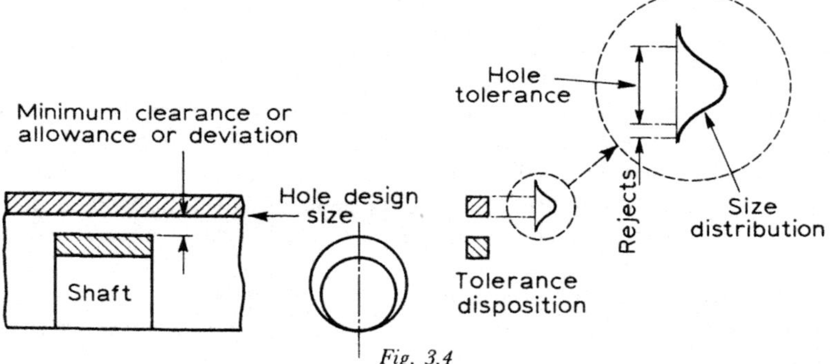

Fig. 3.4

otherwise specified the hole basis (with low limit of hole as the basic size) is accepted as standard, with size variations on a unilateral basis.

One method of accommodating the errors inherent in a manufacturing system is to carry out a geometric analysis covering not only the problems encountered in production and the control of quality but also the assembly and performance of the product. The first objective should be to identify functional features which control manufacture and assembly; remember that the method of dimensioning and tolerancing of such features often implies a preferred machining scheme or operations schedule.

The proper definition of positional tolerances is often overlooked, the maximum metal concept (MMC) or condition, is a principal element in true position dimensioning and implies that an external feature is MMC when its dimensions are maximum and it is perfect in form. A related definition is that of virtual size, meaning the profile generated by the extremes of dimensions controlling position and size. In Fig. 3.5 the space occupied by the hole is controlled by

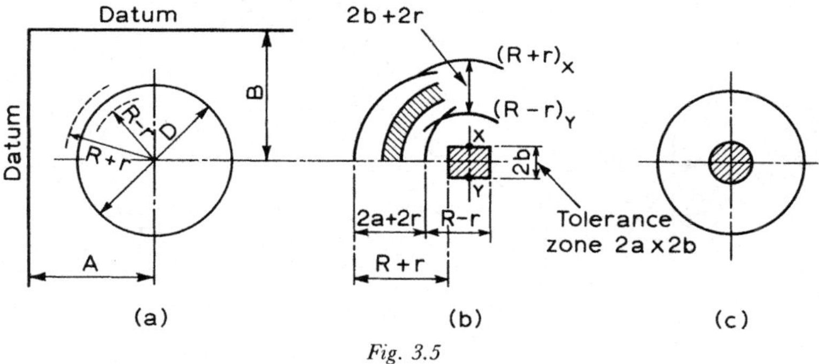

Fig. 3.5

feature dimension D and locating dimensions A, B, on the basis that the edges are a right angle datum. Facets concerned with the application of tolerances to A, B, D, are shown in one quadrant at (b), where any radial pair $(R+r)$, $(R-r)$ within the tolerance envelope will satisfy the requirements; the tolerances are much exaggerated here. A cylindrical virtual hole is provided when true position dimensions are specified for A and B in conjunction with a circular tolerance zone as at (c).

There is much documentation in the literature as to the efficacy of various sampling methods used for inspection purposes. In assessment by attributes each component is regarded as correct or defective, whereas in terms of variables some feature characterising quality is measured by indication. The simplest method of interstage inspection by single samples is to take η objects from a batch N, if δ objects (or less) are defective the batch is accepted, otherwise it is rejected. Some purchasers specify the maximum fraction defective δ they will allow, and for a small sample, when cost of sample inspection is neglected, the cost of quality per unit is:

(i) $\delta(C_R + C_D)$ if batch is accepted after sampling.

(ii) $C_I + \bar{\delta}(C_R)$ if whole batch is examined after sample inspection and defects rejected.

(iii) C_R if batch is rejected after sampling.

where C_R is replacement cost,

C_D is defects cost,

C_I is cost of inspection for batch.

In the typical case shown in Fig. 3.6 where

$$1/C_I > (1/C_R) + (1/C_D)$$

the cheapest method for $\bar{\delta} < \bar{\delta}_A$ is to accept the batch after sampling.

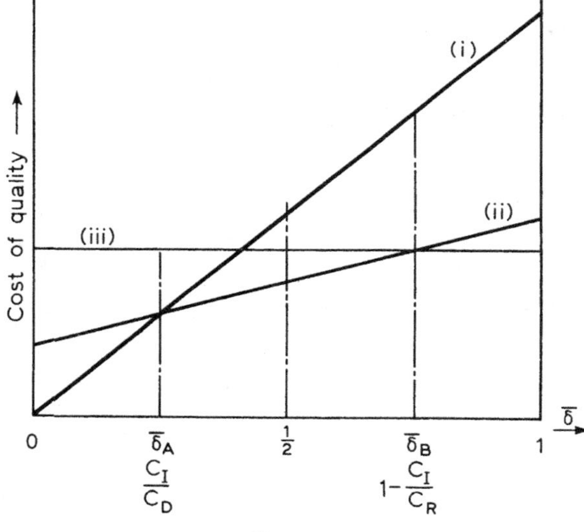

Fig. 3.6

If $\bar{\delta}_A < \bar{\delta} < \bar{\delta}_B$ the batch should be fully inspected; or rejected immediately when $\bar{\delta} > \bar{\delta}_B$.

Even when acceptable size variations have been agreed, some allowance needs to be made for departure of elements from their theoretical geometry as such errors affect both performance and life. A surface scratch might form the site of an area of stress concentration and the stress variation which results may be significant if the load is cyclic.

The concept of notch sensitivity suggests

$$q = 1/[1 + (a/r)]$$

where r pertains to scratch dimensions and a is a function of grain size. Taking values of 0·005 and 0·01 respectively gives $q = 0·048$ and substituting in the Peterson expression

$$K_F = 1 + q(K_T - 1)$$

for a geometrical concentration factor K_T of 4 gives K_F 1·144. Comparison of this value for various materials (all of which may have the same K_T) will reveal its significance and may indicate the need for further machining to remove the scratch or minimise the value of the stress concentration multiplier. Thus any assumption that variations in shape, geometry and finish are limited by tolerance to small proportions is dangerous except in clearly defined static situations.

The reliability of a material or a product is often based on statistical evaluations of acceptable risks and it is generally agreed that a spread of performance in terms of working life is a better concept than a lower level of acceptance. Reliability is partly reflected as expense in assessment of quality or in development testing and the choosing of a low standard is invariably reflected in increased maintenance and replacement costs.

3.4. THE ROLE OF Q.R.

This definition of quality and reliability must be read in terms of a failure concept as applying not only to a service break-down but also to a lack of success in producing a product at a price which will facilitate its sale. In striking a balance between quality and cost regard the former as a combination of aptness for purpose and a measure of consistency or dependability; remembering of course that cost may demand the specification of ultra-smooth surfaces or decorative coatings being restricted to situations needing sales appeal stimulation. Such refinements demand additional processing which may appear to make little economic sense in heavy engineering where outward appearance is often regarded as secondary to function. The topic of form design is closely associated with these aspects and is considered further in Chapter 6.

Initial enquiries as to the relevance of the Q.R. function must include the assessment of factors such as working environment and the maintenance facilities available, since these will to some extent influence the choice of material and process and may demand the availability of additional test/inspection facilities. Recent developments in non-destructive testing techniques give the manufacturer a much better opportunity of ensuring that reliability comes by design and not by chance. The use of penetrant and magnetic methods, radiography and ultrasonics have made a considerable impact in the field of capital goods where mere size negates the use of conventional tests to destruction. It is necessary of course to justify these expensive tests in relation to the value of the component and to the risks involved if a failure occurs in service.

EXERCISES

1. Discuss the necessity for inclusion of some classifying scheme for grade of quality (indicating degree of manufacturing difficulty) in any viable system of limits and fits.
2. Show how assessment of the combination of positional and size tolerances is used to determine the size of a series of clearance holes fitting over a group of fixed bolts.
3. Review the role of geometric analysis in assessing the ease of assembly for a series of components manufactured under controlled conditions.
4. Explain how probability theories may be used to control manufacturing processes and how significant increases in feature tolerances may be permitted without unduly increasing the % rejects.
5. Discuss the essential relationships between function, size and cost in a logical system of material selection.
6. Explain how situations leading to the formation of stress raisers arise during manufacture by casting and machining processes, and how their effect on stress is assessed.

4

CASTING TECHNOLOGY AND POWDER METALLURGY

Production processes may be classified in terms of:
1. Forming from a liquid or a particle state, which is the subject of this chapter.
2. Forming in the semi-solid state.
3. Forming by removal of unwanted material.

The primary object of forming processes is to produce the desired shape with a satisfactory surface finish. Particular specifications for hardness and grain structure, although expensive to achieve, are of a secondary nature. The many alternative casting processes will be recognised as typical of formation from a liquid state and the choice between such processes is invariably economic. Thus, knowledge of physical size or weight is not sufficient for a true comparison as related parameters such as dimensional tolerance, surface finish, minimum section thickness, batch number, cost of labour and capital equipment have to be taken into account. At one end of the scale is sand casting which might have a competitive size range up to 111 N per piece in ferrous materials for unit and small batch production, whereas die casting of aluminium in blanks weighing a few ounces might require a batch of 1 000 to be economic.

The optimisation of casting design for maximum strength and convenience of process should be a flexible procedure. A typical example might be a stress reduction of one-third at edge of boss obtained by altering the cross-section of the lever shown in Fig. 4.1, although this improvement needs to be evaluated in terms of economy if coring and pouring arrangements are made more difficult.

A sound basis is to start with minimum possible section thickness and the basic tapering/enlarging concept towards the source of

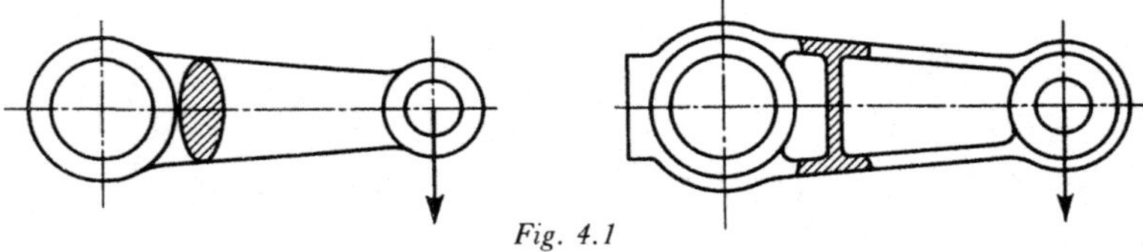

Fig. 4.1

metal feed. Then the casting design can be simplified in terms of joints, partings, cores, etc. Typical joint proportions are drawn to scale in Fig. 4.2.

4.1. CASTING PRINCIPLES

Consider now the various processes involved in the production of sound castings. In mould design, knowledge is required of heat transfer during solidification and of flow patterns, so that suitable risers and a system of channels or gates can be introduced. These will vary with the material as a system suitable for steel would probably cause turbulence such as would oxidise an aluminium

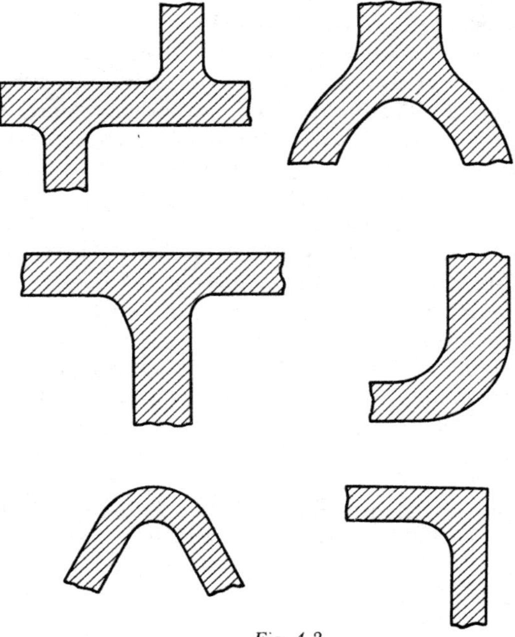

Fig. 4.2

casting. Undesirable thermal gradients in pouring and solidification can be reduced by attention to mould shape. Control of melting, refining and pouring of the liquid metal involves the principles of metallurgical thermodynamics to solve problems such as the precipitation of bubbles, due to gases in solution, which may lead to

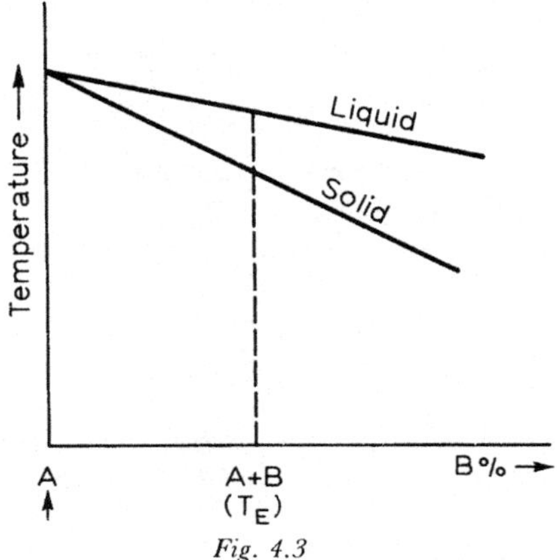

Fig. 4.3

holes during solidification. Fluidity in cast metals does not conform to the chemistry definition of a viscosity reciprocal but is rather a quantity which permits evaluation of the ability to fill a given type of mould at a particular temperature. In the solidification of alloys as opposed to pure metals we need to think of a freezing range, and that composition of solid differs from liquid and there may be more than one solid phase crystallising. Fig. 4.3 depicts a typical AB

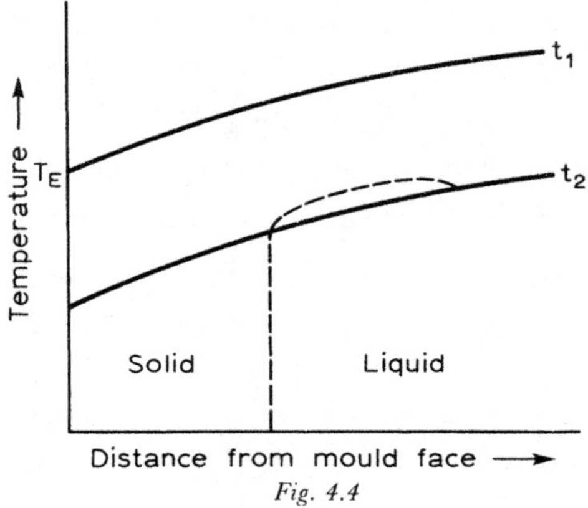

Fig. 4.4

mixture where addition of the latter depresses the freezing point of the liquid A. During solidification the B atoms in A structure are in solid solution. Effect of position in mould is shown in Fig. 4.4, if crystallisation begins at temperature t_1 the difference at t_2 between the alloy and the pure metal is that the temperature of the remaining liquid varies with the distance from mould interface.

4.2. CASTING METALLURGY

Turning now to problems found in real castings, a study of the iron-carbon system will help in understanding the process of casting solidification. In a melt of pure iron crystallisation begins at 1539°C (2803°F), whereas in a solid solution 0·6% carbon the first crystals start to grow at 1482°C (2700°F), while in a entectic solution 4·3% carbon the liquid cools to 1129°C (2065°F) before solidification begins.

Fig. 4.5 shows the micro-structure of grey iron with the conventional pearlitic and ferritic constructions, typical of moderate

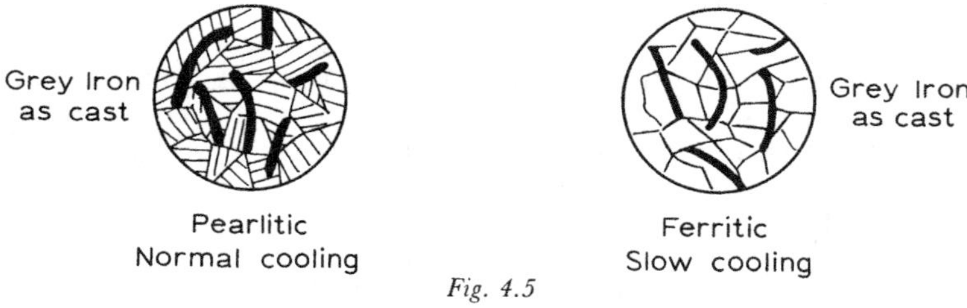

Fig. 4.5

and slow cooling respectively. The presence of carbon in this flake-like form breaks up the continuity of the structure, the material is weak in tension and has poor ductility characteristics. If, however, suitable alloying is arranged, the carbon is caused to appear in a spherical or nodular form and the ductile versions of the pearlitic and ferritic constructions are shown in Fig. 4.6. The freezing state

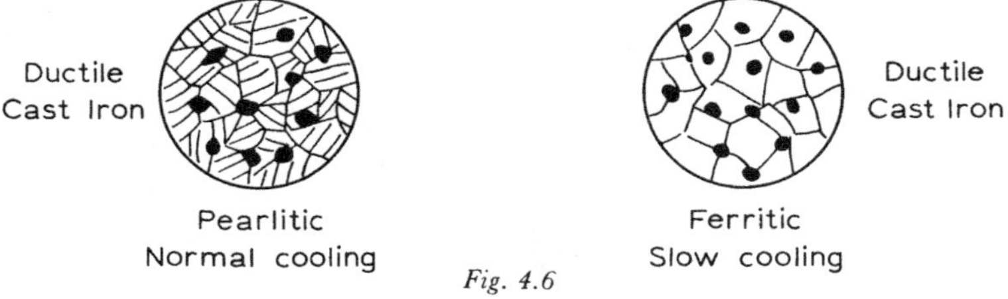

Fig. 4.6

does not proceed along a smoothly advancing front. Fig. 4.7, representing a small casting, shows that crystallisation at the centre occurs several minutes prior to complete solidification.

The effect of mould material on the formation of dendrite patterns needs to be considered in terms of improving the final

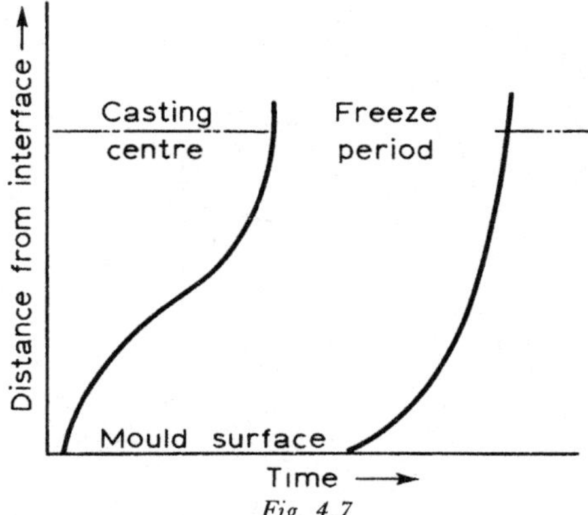

Fig. 4.7

stages of pouring. The risering arrangements depend on the freezing pattern and are organised to prevent the formation of shrinkage voids. For a plain carbon steel the contraction in liquid metal might be 1·6% by volume per 38°C (100°F), probably double this figure in liquid to solid contraction and 4·5 times it by volume during solid state cooling to ambient temperature.

Another major requirement is a gating system to allow distribution at a convenient rate without excessive temperature loss, free from turbulence and entrapped gases. A simple equilibrium expression such as that due to *Bernoulli* can be applied to the ideal situation of an incompressible fluid poured into cavity enclosed by impermeable walls, although the real situation in sand castings is a permeable mould. Considerable thought should be given to the position of metal inlet and whether the pouring arrangement should be horizontal or vertical.

A useful plot for investigation of casting properties is the stress/strain relationship at various temperatures as shown in Fig. 4.8. The differences in contraction of mould and casting and the time difference of contraction at various locations are common stress raisers leading to hot tears, a serious defect which may occur between the start of coherence and the end of freezing. Metal moulds are usual when a quick solidification is desired, typically for ingot casting and many steel ingots are de-oxidised (killed) if being prepared for later forging, to reduce the risk of blow holes.

A requirement for large numbers of small weight castings in light alloys should cause the designer to think in terms of precision casting processes. The term die casting (or permanent mould casting) as used in this country means gravity pouring into metallic dies but in the U.S.A. it infers the use of pressure to force in the

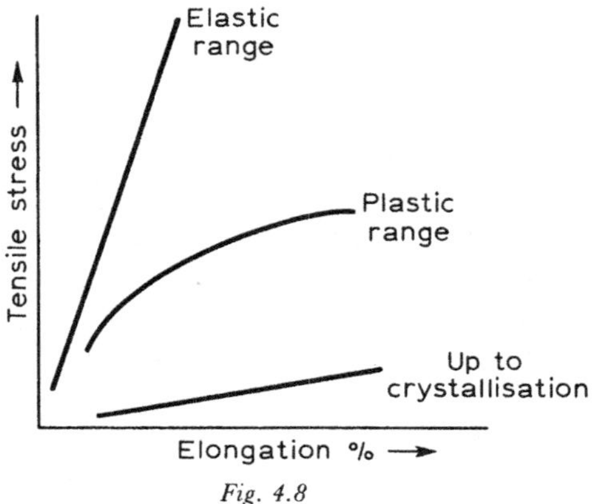

Fig. 4.8

molten metal. The process is considered in some detail in Chapter 9 where non-ferrous metals are dealt with.

Precision investment casting (the lost wax process) is much favoured for small piece production and a particular advantage is the considerable reduction in subsequent machining which needs to be performed. The investment material poured into the wax pattern is usually a refractory aggregate, such as powdered silica with a suitable binding. It is rare that piece weights will exceed 22 N and they are often much smaller than this.

4.3. POWDER METALLURGY

The technique of manufacturing from a particle is commonly called powder metallurgy and involves the heating of a powder, which may have been previously pressed in a mould to ensure cohesion. The object of heating is to cause diffusion and grain growth, during which it may be necessary to protect against oxidation. Amongst the techniques of manufacturing metal powders can be listed:

1. Chemical reaction; which at low temperatures means by the use of a reducing re-agent having a high affinity for the metal it is desired to remove. Alternatively a compound can be decomposed into its elements if exposed to a sufficiently high

temperature. Many theoretically feasible processes are not economic, the method of reduction of oxides permits close control over particle size.

2. Electrolytic deposition; it is not difficult to cause a metal to precipitate on a cathode in a form from which powder can be obtained. Typical usage of aqueous electrolytes and liquid metal cathodes involves further treatment since such powders are generally reactive.

3. Powdering of solid materials; there is limited usage on the grounds of cheapness or for grinding of rare metals in the absence of oxygen. Pulverisation of scrap metal is limited by the problem of holding it and the relatively coarse product may be an expensive comminution, even if the original process was cheap.

Metal powders for industrial components are required to exhibit specific characteristics and are grouped as *dense* (minimum porosity as in tools) or *porous* (as in bearing materials); good compressibility implies minimum cold work in preparation and minimum carbon content, which restricts the choice of reducing agent, particularly since impurity level must be kept low.

The sintering process is based on the adherence of metal powder particles to each other under the influence of heat, until the loose mass of powder becomes a porous solid. The process involves change in particle shape, usually with an increase in density, irrespective of initial compression. In the practical sense interest lies in the rate and completeness of densification. Prospects of adhesion between internal surfaces improve with the fineness of powder and the effectiveness of deformation. The reactions between solid particles are structure sensitive and must conform to shape and orientation of crystals or to lattice defects, similarly account must be taken of surface films or a gaseous/liquid phase occurring during the sintering process.

The limitations of structures obtained by conventional methods lies in their having to comply with an equilibrium phase rule and in alloys, solubility is necessary in the liquid state, but in materials produced by powder techniques various mixtures can be compacted and sintered including non-metallics, although there is no direct relationship between hardness and tensile strength. For such materials basic strength and toughness are of smaller scale and careful consideration needs to be given to choice of test to demonstrate function.

The problems involved in pressing a powder need consideration of welding or adhering of particles, simple applications of mechanics arise in measurement of adhesion force which is related to atomic bonds and chemical affinity. In few cases is a solid pore free object required, in the usual product the ratio of volume of powder of solid metal is less than 3/1 and there is volume porosity of less than 65%.

The real area of contact is an important parameter since the roughness of the prepared surfaces may exceed the range of atomic attraction. Neither can we assume that friction is entirely due to adhesion as some restacking of particles occurs when pressure is applied, which may invalidate theoretical pressure/density relationships. Oxide films will similarly exert a marked influence on frictional behaviour. Die friction should not be overlooked; reaction is determined by the amount of cold working between powder and die walls. The usual assumption of pressure loss proportionality with die wall area involves some generalisation in assumptions of density distribution. There is need for careful consideration as to the employment of a lubricant to reduce the power required as mixing of lubricant and powder will reduce density and rate of flow. The strength of the cold-pressed compact is thus reduced and undesirable internal stresses may result and above all there is the problem of removing lubricant before sintering.

The manufacture of cemented carbides for cutting tools or dies is a common application, after preparation of say, tungsten carbide and cobalt, to particle size, pressure compacting is carried out. This is followed by sintering in a controlled atmosphere until components are strong enough for shaping; the final sintering may be above the eutectic temperature with liquid infiltration assisting in production of a denser structure.

EXERCISES

1. List and describe the general nature of phases and constituents found in an iron-iron carbide alloy system.
2. Compare the solidification of a cast alloy with that of a pure metal in terms of microstructure for various cooling rates.
3. Compare three alternative types of casting process indicating how a choice would be made for a particular application in terms of the economics involved.
4. Suggest a typical component to be produced by investment casting and outline the manufacturing procedure.
5. Discuss basic factors relevent to the choice between ferrous and non-ferrous metals for casting of consumer goods.
6. Discuss the relationship between atomic bond, chemical affinity and adhesive force in a component made by pressing particles.

7. Describe the phases in the process of producing metal powder parts and name three components which could not be produced by other methods.
8. Explain problems encountered in the Sintering process when a lubricant has been used to reduce power requirements in the pressing phase.

5

PROCESSES USED IN FORMING FROM A SEMI-SOLID STATE

It is well known that the effect of plastically working a metal is to confer strength and increase resistance to further working. Such deformation processes perform a useful function, since a cast structure composed of large, low strength crystals is the reverse of what is required in a load carrying component. The loss of ductility associated with work hardening has superimposed effects due to the recovery associated with recrystallisation at higher temperatures. If the process is finished early, at temperatures above recrystallisation, the grain growth associated with cooling may be a problem.

In representing forming processes on a stress/strain curve it should be realised that the straining range in metal working may be one hundred times more than in the elastic range, so that assessment of change in shape concentrates on plasticity effects and assumes no change in volume due to plastic flow. In terms of semi-solid state forming we can classify initially on the basis of: forging, rolling, pressing, extrusion, drawing. These processes will be described later.

Size alteration during a process is used to classify large or small scale deformations, with sub-divisions in terms of hot and cold working, the latter meaning a range of progressive hardening as plastic deformation increases.

5.1. WORKING TEMPERATURE RANGES

Common metals have distinct temperature ranges for hot and cold working, Fig. 5.1 is typical although ranges are difficult to calculate because of impurities. Almost all resistance to fracture is lost at the semi-solid/liquid changeover and the working range may finish

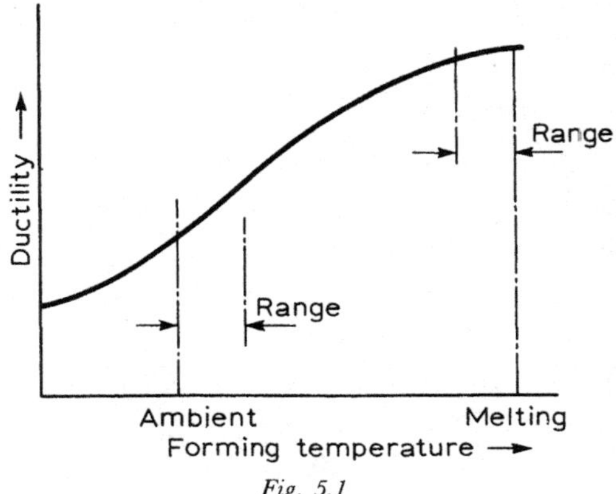

Fig. 5.1

38°C below the theoretical melting point. In terms of deforming a certain volume, the effect of temperature on force is much simpler in character than when ductility is considered; flow stress decreases with temperature increase but its rate of change depends upon the speed of forming, Fig. 5.2 shows typical variations.

Speed has considerable effect on metal performance during processing and it is necessary to ensure that acceleration can be con-

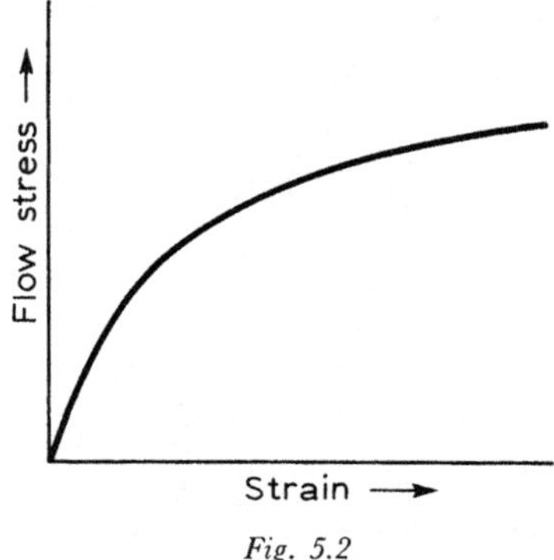

Fig. 5.2

trolled in the primary plastic period where stress distribution is difficult to predict. In certain temperature ranges a metal may evidence brittle behaviour at low speed but satisfactory ductility at higher speeds, Fig. 5.3 is typical for such behaviour. The pronounced difference between slowly applied loads and the impact effect of metal working processes is particularly important in the early stages of ingot transformation; voids, porosities and inclusions

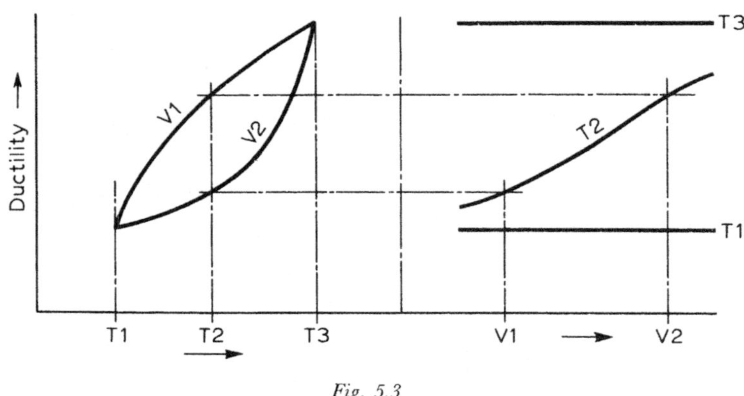

Fig. 5.3

responsible for low strength areas can be ameliorated by rapid working and interstage soaking. Most pure metals and solid solutions are hotworked at temperatures where their ductility is considerable. Fig. 5.4 shows the effect of alloying as a displacement of the lower limit of the hotworking range.

Mechanical directionality refers to the property variations which

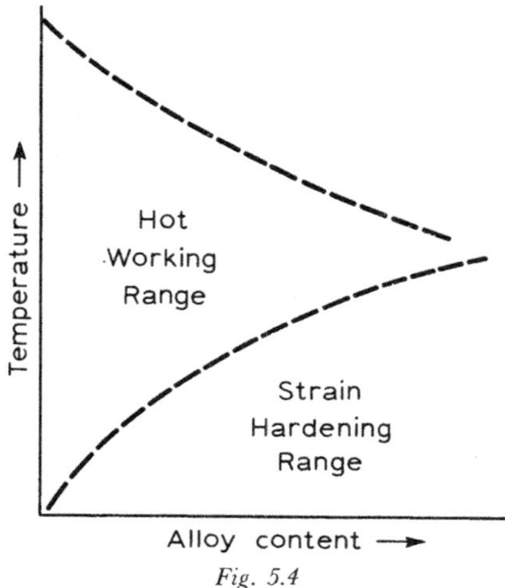

Fig. 5.4

occur in wrought products and Fig. 5.5 shows the effect of the processing direction with progressive improvement in longitudinal directionality. Ductility is sometimes sensitive to the forming method as some processes 'tire' or exhaust the material more than others. Fracturing and necking are common types of forming failure, the latter is analogous to a tensile test situation and is a measure of stretching ability. Striations showing metal flow permit assessment of the possibility of fibre separation which limits any subsequent cold working sequence–even when heat treatment follows the initial forming process there may be considerable distortion if stresses run counter to fibre direction.

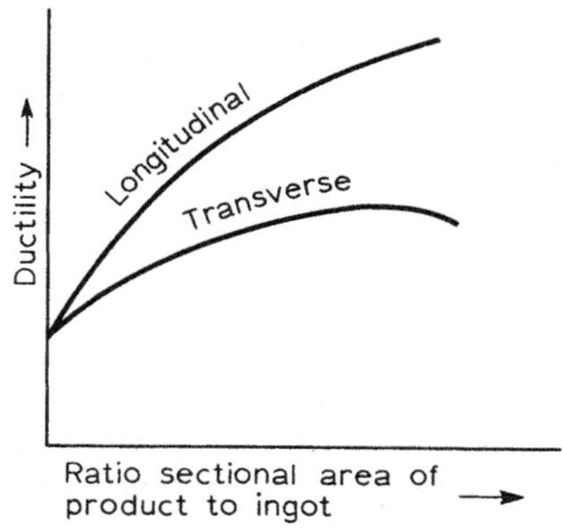

Fig. 5.5

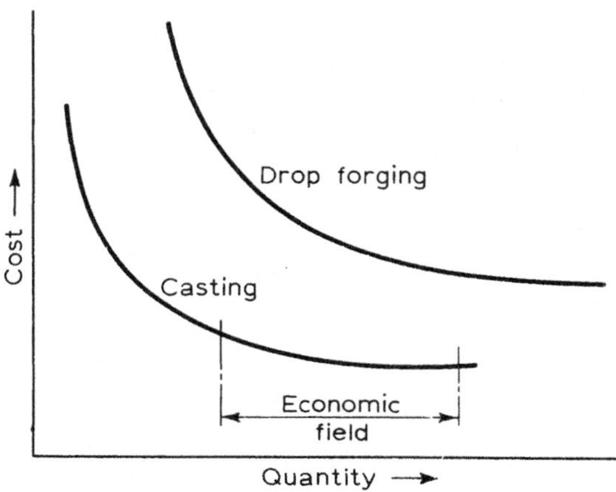

Fig. 5.6

Direct compressive situations are found in forging, rolling and extrusion, they have similar predisposition to failure but the predominantly compressive arrangement means that the elongation typical of a tensile test can be exceeded many times during forming. The tension case is characterised by stretch forming and that of indirect compression by *drawing*, tool pressure is much lower than in the direct method but large stress variations occur due to tension at certain situations. For ductile materials 'drawability' is a sensible forming limit.

Criterion in the *bending* process is the smallest radius to avoid edge cracking, if the optimum value is exceeded there is spring back from tool shape and a non uniform state of stress is found at the deformed section. Assuming availability of material, the quantity required determines the economics of process selection, Fig. 5.6 shows comparison between drop forging and casting; fabrication by welding would be cheaper than forging for any quantity.

5.2. FORGING PROCEDURES

The problem in forging is to utilise techniques to produce the fibre rearrangements leading to toughness and impact strength which differentiates forged materials from those produced by other methods. Thus, an empirical deformation requirement might be a cross section reduction of at least 80% to ensure that original grains are broken down. The drawing and widening which occurs will compact the structure, reduce grain size, increase strength and produce fibre flow in the direction that strength is required.

Much large-scale forging is performed hot and die friction may be large enough to raise transverse shear stress to a value approach-

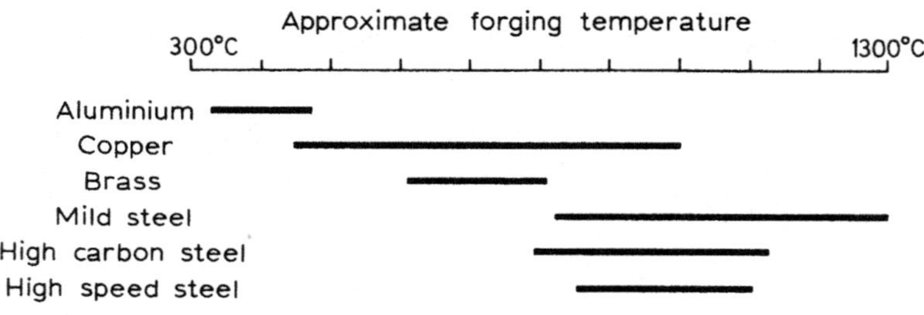

Fig. 5.7

ing yield. A forging temperature range chart such as Fig. 5.7 is a useful guide. When cross-section is large, hammer forging may not sufficiently work the core, whose structure may be left in a coarse condition, so that press forging techniques have to be introduced. Attempts to explain forging theory on the basis that plane sections remain plane have met with moderate success but interface friction is difficult to predict.

Form design of forgings is influenced by the provision of draw or draft so that the piece may be withdrawn from the dies. In drop forging, sectional variations demand careful die design and problems such as flange chilling require particular arrangements for shape or disposal of metal. In the up setting process there are limit-

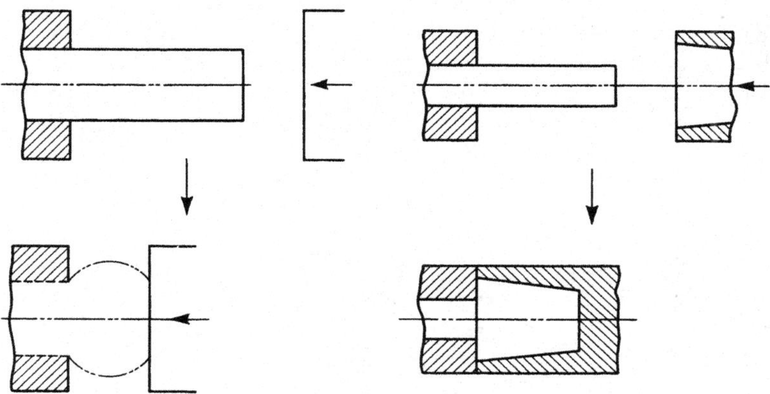

Fig. 5.8

ations on unsupported length in terms of diameter and die design must provide the necessary support, see Fig. 5.8.

Cogging is a part finish process to prepare pieces for later rolling; there is considerable variation, in load due to changing area of contact and material properties. Fig. 5.9 relates variables such as

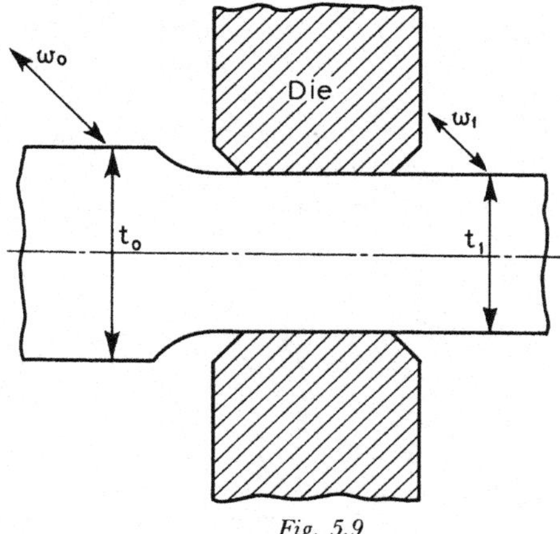

Fig. 5.9

squeeze $t_0 - t_1$,
squeeze ratio t_0/t_1,
spread ratio w_0/w_1.

5.3. HOT ROLLING

Rolling functions include forming of an ingot or cogged piece for later small-scale processes and the conferring of improved properties. In the simplest analysis shown in Fig. 5.10, the roll separating force

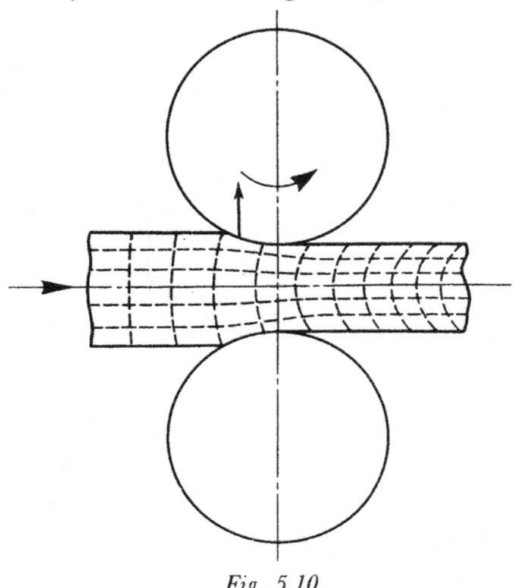

Fig. 5.10

is considered to act at the point of no slip where material is moving at same speed as rolls. As the friction coefficient increases, the neutral point moves towards roll gap and this value determines the amount of reduction possible.

A situation can be reached where no further reduction is possible, roll gap has opened due to distortion from the force trying to produce strain on a thin piece of hard metal. Separating force is reduced

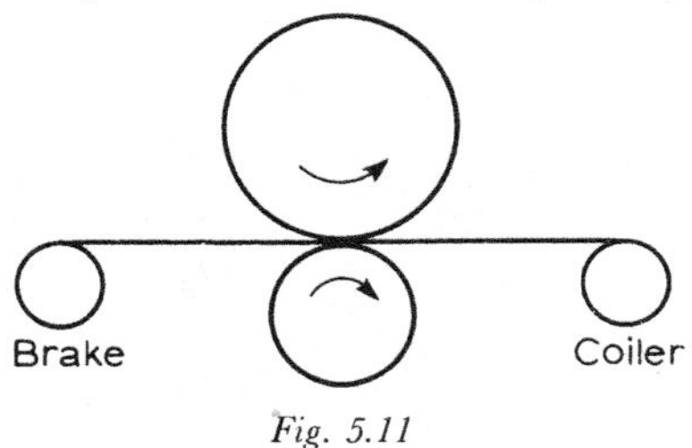

Fig. 5.11

by smaller rolls or by application of tension about the roll gap as shown in Fig. 5.11.

There are many problems in the implementation of hot rolling theory quite apart from the techniques of pass design. Fig. 5.12 shows a typical scheme with ● representing a particular bar element

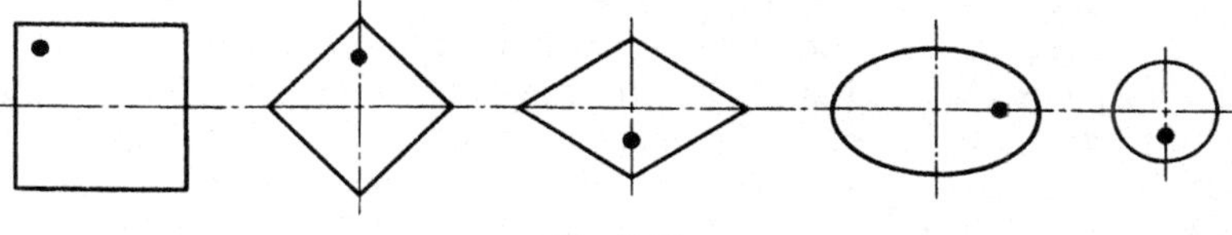

Fig. 5.12

at each stage. Thus, dependence of yield on several parameters, poorly defined frictional situation, plain strain deformation, means that the number of assumptions permits only approximation to values of roll force and torque.

5.4. PROBLEMS ENCOUNTERED IN COLD WORKING

It is appropriate here to consider the main differences between cold working and the temperature-dependent processes dealt with earlier. The effect of grain size and shape is overshadowed by porosity and segregation in a hot process but, when the material is cold worked, grain shape is affected by piece shape, quite apart from the fundamental properties of malleability and ductility which cold working demands. When cold working is specified as a finishing process, intermediate heat treatment will be necessary; properties such as malleability, ductility and toughness will determine material suitability.

The problem of crystal distortion and grain size is paramount in the work-hardened situations typical of cold working; capacity for further work is dependent on the annealing or normalising provided as an interstage treatment, only in this way can close tolerance and good finish be maintained. Cold pressing makes particular demands on material properties such as ambient plasticity and the pressing operation, whose use infers quantities sufficient to warrant tool costs, involves such widely varying processes as blanking, piercing, cupping, bending, etc.

5.5 EXTRUSION PROCEDURES

The basis of the *extrusion* process is the conversion of a billet to a length of uniform cross-section as it passes through an orifice. The output required must be sufficient to justify the high capital cost. In the direct process the metal is forced towards a die at one end of container, whilst in the inverted or indirect process, the die is placed in a ram which is bored to allow passage of extruded bar, this is less demanding on power and product is better worked. Properties are influenced by type of flow, chilling of surface due to contact with the die means that the core is more easily extruded. Skin effect must be considered in any assumption of a streamline concept since this depends on friction effect and the non uniform material plasticity, together with operating speed. Fig. 5.13 shows pressure variations at optimum speed.

The hot working range is from hot-shortness to over-stiffness. Work must be performed while the metal has sufficient plasticity to

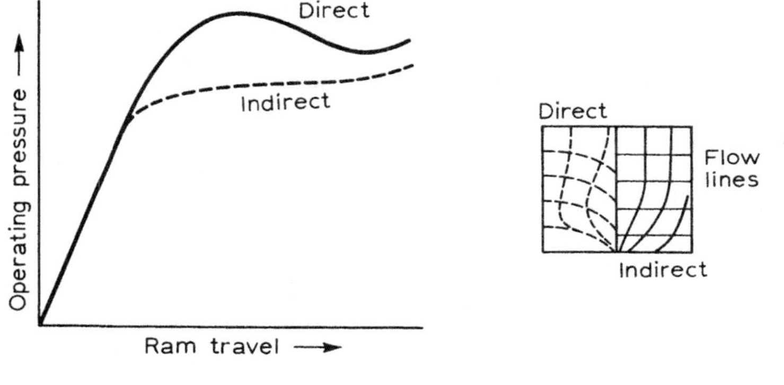

Fig. 5.13

allow shaping within the power which is available. An extrusion area ratio of at least 7:1 is commonly specified to ensure sufficient working of structure. For a given extrusion pressure the extrusion ratio obtainable increases with temperature up to the point of hot shortness. A chart such as shown in Fig. 5.14 can be drawn up to define the allowable extrusion range, as the region under a curve of constant pressure and a given speed.

Much recent work has concentrated on the development of hydrostatic extrusion where the billet is extruded through the orifice by fluid which is pressurised by the stem. Pressure requirements are much reduced (20% is not uncommon) since there is no container wall friction which conventionally limits billet length, the fluid assists in lubrication.

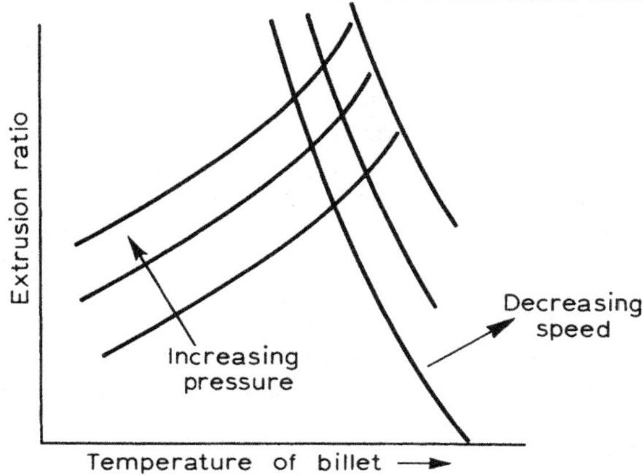

Fig. 5.14

5.6. PRESSING AND DRAWING

The term *pressing* has at times been applied to various forming operations including *stamping, piercing, blanking, bending* and *drawing*; the latter is an indirect compression process. The effect of the tensile applied forces is less than the reactive compressive stresses and the combined action producing metal flow results in tool forces

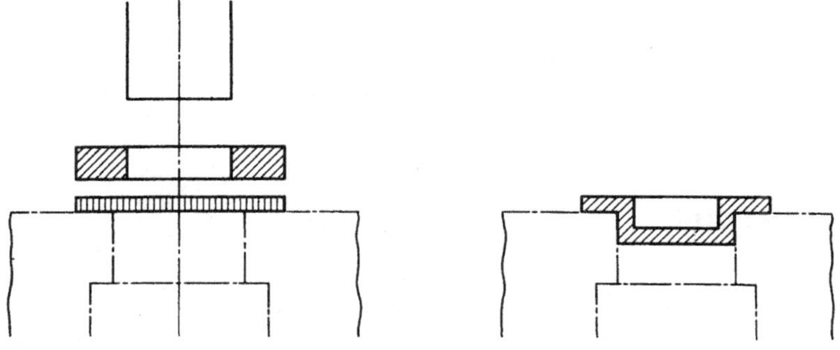

Fig. 5.15

lower than in direct compressive operations, although there is considerable stress variation in the vicinity of location points.

Cold working is common because of property improvements and finish desired but demands considerable ductility; cross-section reductions over 30% in mild steel are not uncommon in dry drawing.

Much bar stock is required with a bright finish, the drawing process is convenient to ensure accuracy and repeatability of profile. In rod or wire drawing the applied force creates tension at the exit which is supplemented by radial and circumferential compression at the die and this is the condition of maximum size reduction. If a back-pull is provided to improve die life, per cent reduction is decreased. The hydrostatic method is also being applied to drawing techniques and has greatly increased the scope of single pass reductions.

Deep drawing from sheet material involves initially forming a cup shape as shown in Fig. 5.15 and should be regarded as a distortion process involving two-dimensional plastic strain. With clamped edges, the system is essentially tensile and material tends to become thinner, with free edges the diametral shrinkage tends to prevent thickness decrease while increasing it in radial direction. Force prediction is mainly empirical as the depth of single draw is based on tensile strength and thickness and many assumptions are made as to property relationships in the plastic region. Hardening due to cold working imposes restriction on amount of draw before annealing and redrawing is necessary above 60% cross-section

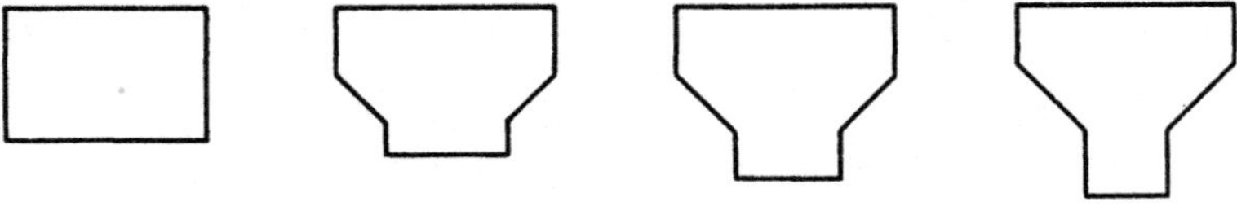

Fig. 5.16

reduction or for a similar single reduction of a blank to form a plain cup. Fig. 5.16 shows how successive draws will produce a desired shape from a cup blank using reducing dies.

EXERCISES

1. Review typical applications for which the properties of a forging would be considered superior to those of a cast product.
2. Discuss reasons for the difference between the forgeability of pure metals and alloys.
3. Calculate the ideal work of deformation in a forging process, relating the reduction per blow to the forging velocity. How can a true compression stress/strain curve be utilised?
4. List parameters concerned with die forging processes necessary to express the change of shape in terms of the coefficient of spread.
5. What properties of metals are most important in the estimation of rolling loads for both hot and cold processes?

6. Define the ratios involved in the geometry of the rolling process and indicate which of these are independent of the scale of the operation.
7. Assuming an empirical expression for mean rolling pressure, plot curves for variation of rolling load with diameter, coefficient of friction and thickness.
8. Explain the effect of deformation on grain structure for both hot and cold extrusion processes and list factors which determine the choice of process.
9. Derive an expression for the ideal work of deformation in extrusion and explain the effect of lubrication on flow patterns in direct extrusion.
10. Assuming that billet dimensions and final size of an extruded bar are given, deduce whether it would be more economical (in terms of pressure) to increase billet length or its diameter, if the length of the product is to be doubled.
11. Derive an expression for the ideal work of deformation per unit volume in wire drawing.
12. Sketch typical curves of draw to yield stress ratio/reduction in area for all practical parameters in rod and wire drawing.
13. Explain the procedure of tests commonly undertaken to assess the behaviour of sheet metal during deep drawing operations. Under what circumstances is lubrication considered essential?
14. Discuss the state of stress in cup drawing at edge of blank, under punch and in cup wall.
15. Discuss the relationship between hot working, cold working and re-crystallisation of ferrous materials for all the processes considered in this chapter.
16. Explain how the fatigue resistance of a metal can be improved by cold working and how subsequent operations are affected by residual stress.

6

ECONOMICS OF FORMING BY MACHINING PROCESSES

It is not the intention in this chapter to launch into a numerical review of feeds and speeds or the elements of machining techniques which are well described for various operations in technician's handbooks. The considerations will be in terms of the principles of forming by machining, the process economics involved and how form design must be taken into account by the product designer.

6.1. KINEMATIC CONSIDERATIONS

Provision of a particular shape involves analysis of the combined kinematic movements of a machine tool in terms of freedom and constraint. Such a combination of movements results in a generating process, while a forming process results when the work shape is a copy of the tool surface. Flat surfaces are produced by a line advancing in its own plane. The method must be capable of achieving the required accuracy in terms of flatness and relationship with adjacent planes.

Typically such surfaces are produced by reciprocating movements, by facing in a lathe, by face or end or slab milling. A cylinder is produced as a straight line rotates about a parallel axis or as a circle moves in a line parallel to its plane, the result of turning, boring, drilling, etc. Kinematic requirements thus depend on the combination of rotation and translation; for each of these, the motion involves five points of location and one degree of freedom – the rotation of spindle and the translation of saddle on a lathe are a typical example.

The product designer must first consider the rules which govern manufacture by machining. Competitive manufacture requires that machining work be kept to a minimum and this demands appropriate form design, not only for the desired shape but also to provide access for handling and holding. Additional design considerations might involve arrangements for the inclusion of standardised accessories, requirements for planned maintenance, etc. Thus, the first effect of applying economic considerations to metal removal operations is their influence on earlier decisions as to form, shape and material. An optimum situation can be defined for each combination of variables in metal cutting. Typically, time based cost decreases with an increase of speed and feed, although such alterations must also effect cost in terms of shorter tool life. Limits on feed are prescribed by maximum tool force for which a machine tool is designed. Facets of design related to performance of machine tools are discussed in the next chapter.

6.2. MACHINABILITY

In terms of doing the job in the shortest possible time with best obtainable values in terms of power, tool life and surface finish, a basic definition of machinability would be useful. This must take into account both engineering and metallurgical considerations and it would be convenient if it could be quantitatively expressed in terms of a mechanical property such as strength or hardness, or of parameter measurement associated with metal removal. It is a common error to believe that properties are a direct measure of machinability as additives to improve machining characteristics do not always reveal themselves in simple property tests. Increase of cutting speed leads to decrease in chip thickness and so theoretically to reduction in cutting force, although this force is remarkably constant within practical ranges.

It is convenient therefore to consider specific cutting pressure, variation of which is shown in Fig. 6.1 or a range chart relating speed and hardness for particular materials as in Fig. 6.2.

In practical terms, a machinability chart based on a wide range of feeds and speeds for various material combinations, will take cognisance of the onset of wear. An optimum range might be specified in terms of tool cratering and deformation at one extreme with heavy build up on tool face as the other limit. One facet of the *Brandsma* test, for example, relates performance in terms of the speed at which a reduction of 0·02 mm occurs in the depth of cut due to wear. With certain assumptions as to the orthogonal nature of cutting a linear relationship might be proposed between friction angle, tool rake angle and shear angle, leading to a plot of cutting force and shear force as a function of shear plane area.

Other schemes are based on the machining performance of free cutting mild steel (179 Brinell) which for typical feed, speed, tool life and cutting with high speed steel tools is given an index of 100.

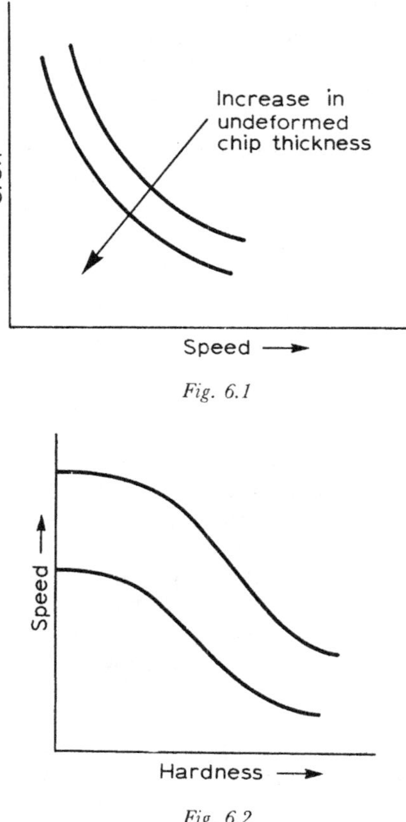

Fig. 6.1

Fig. 6.2

On that basis a 0·4% carbon nickel chrome molybdenum steel (heat treated to 400 Brinell) would have a machinability index about 20. Certain empirical formulae found in manuals of metal cutting, purport to provide a linear equation for the relationship between speed, feed and surface finish. A conservative estimate, not including wear, reveals at least twelve variables in these relationships and such empirical equations should be treated with reserve.

6.3. ECONOMICS OF THE CUTTING PROCESS

Sufficient has been said to indicate that feed and speed are the basis of machining economics and, with these, must go the matter of tool life. The empirical relationship $V.T^X = C$; illustrated in Fig. 6.3 is commonly used to specify an optimum cutting speed where x and C

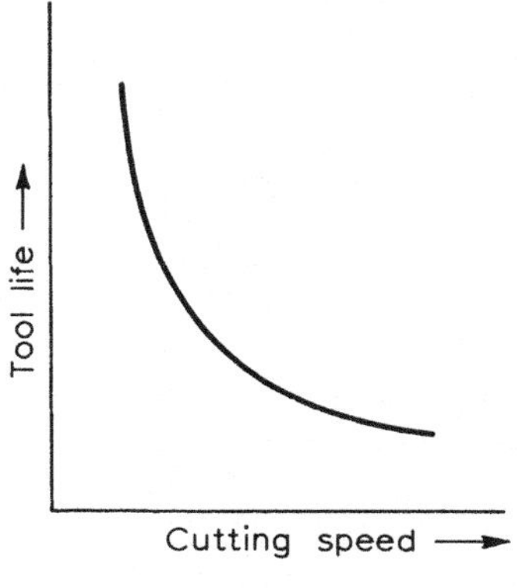

Fig. 6.3

are constants related to tool life considerations. Part of the application problem lies in the correlation between the single point laboratory test and the economics of large scale production. The typical relations which follow emphasize the important trends.

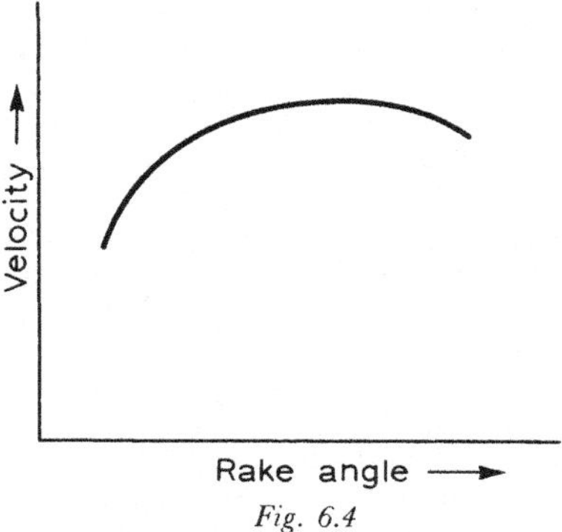

Fig. 6.4

The basic size which affects the wear rate is undeformed chip thickness, speed is conveniently expressed as a single function provided other parameter relationships with thickness are known. One typical relationship, assuming constant tool life, is shown in Fig. 6.4. An increase in feed will always result in smaller life decrease than a proportional speed increase, thus most production engineers think in terms of maximum possible feed before choosing a speed to suit, although this means an increase in cutting force.

A straight-line relationship can be expressed in a logarithmic plot of tool life and cutting temperature and the extension of this to include speed and feed is shown in Fig. 6.5, similarly wear and finish variations with our basic parameters are shown in Fig. 6.6.

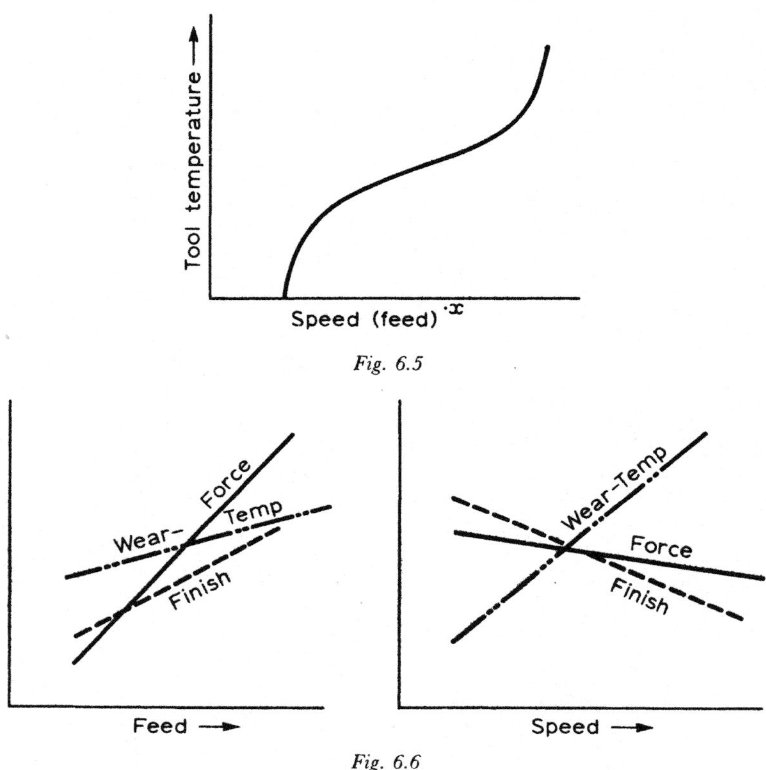

Fig. 6.5

Fig. 6.6

Metal removal processes may affect material properties as a result of a temperature rise in a shearing type of metal removal. Thus, it is sometimes necessary to reconcile the thermal problem and, in particular, the dissipation of heat from the work of deformation in the shear zone. The extremes of the working range involve low speed operation, when there is plenty of time for conduction and tendency to an isothermal state, or at the high speed end of the scale where a non steady adiabatic situation is found.

Summarising the economic parameters, it is apparent that two fields of utilisation for lathework are worth further investigation:

1. In terms of desirable tool life which may demand only a proportion of machine power.
2. In terms of maximum metal removal at full machine power which may result in a relatively short tool life.

The intersection of these regions on a feed-speed plot is shown in Fig. 6.7, which allows comparison of power requirements at constant efficiency. The plot of a series of corresponding power values, see Fig. 6.8, provides an intersection indicating the economical cutting

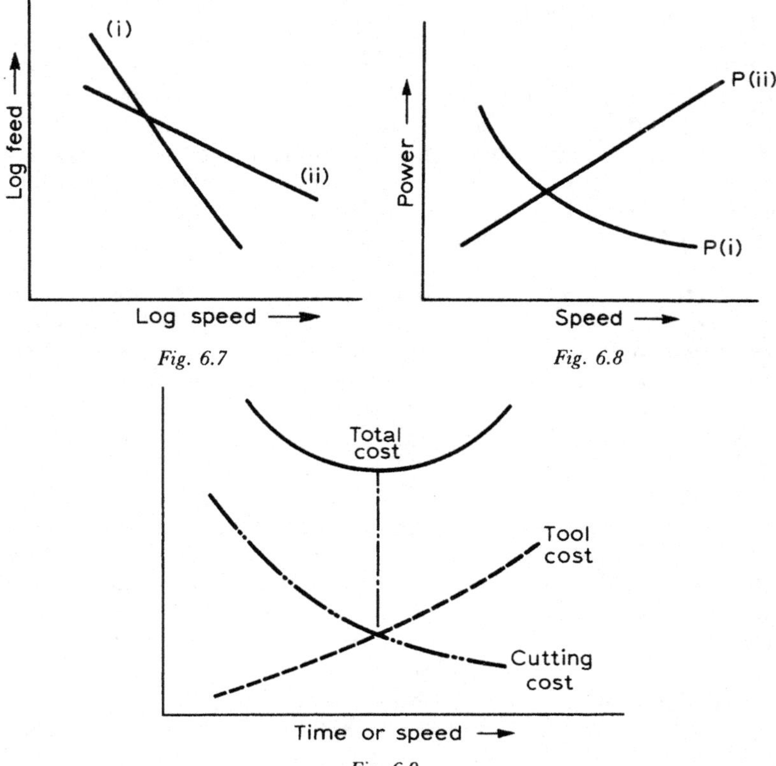

Fig. 6.7   Fig. 6.8

Fig. 6.9

speed which is linked to chip area so that an acceptable feed can be specified.

Workpiece diameter may then be included in terms of a particular speed progression; normally a compromise range is provided reflecting the tool life and economic cutting speed limits already referred to. In general, least working time is found in the region of simultaneous utilisation of life and power compared with maximum life and partial power. The way is now clear for a summation of all the items involved in production costs and the variables presented as shown in Fig. 6.9 can be used to provide an inversion of the total cost curve indicating an optimum choice. For real values allowance must be made for fixed charges, handling time, etc.

6.4. FORM DESIGN

Form design for objects to be produced by machining poses many problems. The product engineer is constantly being invoked to take account of aesthetic and sociological aspects within the framework of available manufacturing methods and the economics of our industrial system. Features such as curves, re-entry shapes and style contrasts can be incorporated without much difficulty in cast or moulded shapes but may involve sophisticated and expensive machining processes if a block of raw material has to be utilised. In drawing up a series of basic rules for manufacture by removal of unwanted material it is assumed that the previous forming process provides a shape of close relation to the final requirement and design for minimum metal removal is axiomatic. Machinability in this context means that the design allows simplified machining processes, space for tool run out might be a typical example or the provision of bosses and access for clamping and checking.

The layout of working drawings will include a clear definition of dimensions and tolerances arranged to suit the most economical machining method.

In considering the wider aspects of form design it is necessary to review the combination of art and science which success in this field demands. If aesthetic principles can be incorporated without any restriction on process cost, the most pleasing effect is obtained when the backbone of a design is a dominant geometric form as any appearance of instability will disturb the effect of balance.

All shapes have a characteristic quality of their own, for example, grouped vertical lines give an appearance of height, but the shape or form of a successful domestic product, while essentially an expression of function, must make an emotive appeal before any question of its technological merit arises. The craftsmanship or technical ability of the industrial designer is not revealed in mere application of formulae but must involve initial visualisation of purpose, scope and construction, before application of engineering considerations in terms of available manufacturing facilities and their effect on cost.

Prime considerations for a domestic appliance involving convenience, attractive appearance and 'failsafe' working are not the major qualities required of a scientific instrument, where reliability, sensitivity and accuracy would take pride of place. Finishing media will heighten the effect of form which the designer has been able to incorporate and changes in gradation of texture and colour will stimulate sight and touch. The swamping effect of large areas can be mitigated just as easily by a change in texture as by a physical division, while developments in chemical and applied finishes have led to considerable improvement in product protection and enhanced appearance.

EXERCISES

1. Discuss parameters involved in the removal of metal under orthogonal machining conditions and show how the forces acting may be represented in a graphical analysis.
2. Discuss the mechanism of chip formation in the machining of plain carbon steels and the effect of common alloying elements on machinability.
3. Review attempts to obtain correlation between microstructure and machinability, indicating the effect of residual stress.
4. Show how the principles of experimentation techniques can be applied to arrange a series of tests involving all the parameters associated with tool life and the effectiveness of cutting fluids.
5. Discuss the compromise involved between stiffness and sensitivity in the measurement of cutting force by dynamometers in turning and milling.
6. Review the emotive, material and commercial aspects involved in product design.
7. Write an explanation of the construction of a simple consumer product (without reference to drawings) describing the various components in terms of their geometric shape.

7

FACTORS AFFECTING MANUFACTURING ACCURACY

The previous chapter discussed the relationships between operating variables associated with metal removal technology. It is appropriate now to consider problems which affect the planner or designer in terms of both component and machine tool characteristics and to examine factors on which the accuracy of the workpiece depends. In 'Engineering Design Problems' (Iliffe) by the same author, Project 17 considered the effect of cutting forces on a machine tool frame. The intention in this chapter is to consider the wider design aspect, including model techniques from which the form design for a cast or fabricated machine tool structure can be evaluated.

7.1. CUTTING FORCES

The basic structure should be sufficiently robust to permit retention of the accuracy 'built in' during its manufacture, such accuracy should not depend on the operator and should be consistently available over the life of the machine. The first requirement is an assessment of tool forces, taking the lathe as example we might refer to the idealised case of continuous chips produced by an orthogonal process. Fig. 7.1 shows the resultant R of a tool force system where P_1 and cutting speed determine power, P_2 is feed force not likely to exceed $\frac{1}{2}P_1$, P_3 in direction of depth setting may be $\frac{1}{3}P_1$. Feed rate and depth of cut determine the chip section and volume of metal removed.

A commonly used parameter is the apparent mean shear strength of the material, based on the production of a continuous chip, but a more important requirement is an accurate computation of the chip thickness. Orthogonal theory permits an estimation of this thickness by considering the chip as a rigid body in equilibrium, other theories are based on plasticity concepts but, in general, there is no unique relationship which agrees with the large mass of

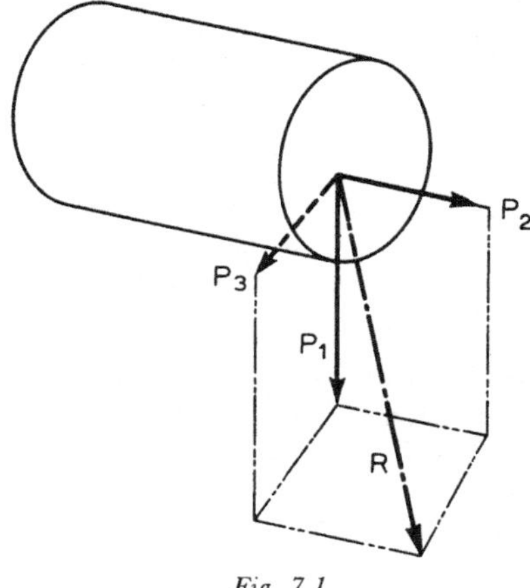

Fig. 7.1

experimental data, probably because of the unresolved influence of frictional behaviour on tool face. In any event the real cutting process involves an oblique situation where the chip flows up the tool face at a particular inclination to a right angle from cutting edge. The correct value of this angle is an important factor in the production of continuous chips and the benefits associated with such a machining situation, but it is not easily organised at speeds below 200 ft/min (1·02 m/s).

Whatever sophisticated theory is used for prediction of tool forces it will avail nothing (in terms of accuracy and fine finish) if the machine tool does not possess the required rigidity and stability.

The basic elements concerned with the geometric analysis of the machining process involve rotation about a fixed position and a matched translation; rotation of lathe spindle and movement of saddle was mentioned as a typical example in the previous chapter. Kinematic systems of freedom and constraint control the movements of workpiece and cutting tool with some linkage to provide a desired relationship between them. Several well-known methods are available for classifying such linkage systems and the geometric shapes which they control and the main consideration is of the surface texture arising from relative geometrical movements between tool and workpiece.

For the simple case of a round-nose tool, deviations may be compared to the series of curves whose pitch is feed, conventionally regarded as the ideal profile. Such evaluations may be in terms of peak to valley height or deviation from an envelope or variation from a mean line, texture will improve as nose radius increases and feed is reduced provided that angle profile is adjusted to suit. Depth of cut is relatively unimportant, provided that there is optimisation of speed and rake angle in terms of chip formation any investigation of surface texture can concentrate on the relation of feed to the other parameters.

7.2. ACCEPTANCE TESTS

The tests originally defined for testing the accuracy of machine tools were on a static basis, primarily concerned with levelling of bed, alignment of the various translations and assessment of the accuracy of rotation. Typical results might be: headstock axis parallel with bed within ± 0.0005 in/ft (0·041 mm/m), leadscrew bearing axis parallel with bed within 0·004 in (0·102 mm), permitted eccentricity of headstock bore 0·0003 in (0·0076 mm); this is a brief selection since up to thirty tests may be required for a conventional centre lathe. In terms of modern requirements for metal removal, dynamic considerations must be taken into account, typically movements which alter chip section and so vary cutting force exercise a significant effect on oscillation during self-excited vibration.

BS. 3800 refers to methods of testing the accuracy of metal cutting machine tools and makes a clear distinction between tolerances for dimension, position and form—the type of geometrical check already referred to involves the interaction of all three types of tolerance.

7.3. FRAME CHARACTERISTICS

As with the many other aspects already discussed, the choice of a machine for a particular duty (assuming that it fulfills basic accuracy requirements) is almost entirely economic. Machine characteristics in terms of standard of performance are always related to the given price and accuracy of the machine elements as reflected in the dimensional accuracy of the components which are produced.

Stiffness, as the ratio of static load to deflection is more important than load capacity, since it is rare in machine tools that element stress is more than a small percentage of the permitted value. Deflection of components diverts the tool from desired path and although in general a combination of bending and torsion will cover the worst static case no real estimate of deformation can be made without allowance for dynamic effects. In bending the force-deflection relationship is proportional to (EI) but stiffness in torsion is affected by material and shape.

Several well-known results can be applied to determination of ideal shape, with deflection increasing as cube of depth it is desirable to utilise the optimum relationship of $(length)^2/depth$ for the stress and deflection permitted in a given material. In large machines the foundation needs to be regarded as an integral part of the assembly; permitted deflection is likely to be of the same order as in smaller machines but the force causing it is much greater so the stiffness problem is more acute, a safe value may not be easily obtained without assistance from foundation.

The flexural deformation of a bed and its foundation can be assessed by the conventional differential equation, but the complex situation produced by levelling pads requires assumptions as to ground interaction before the moment diagram depicted in Fig. 7.2

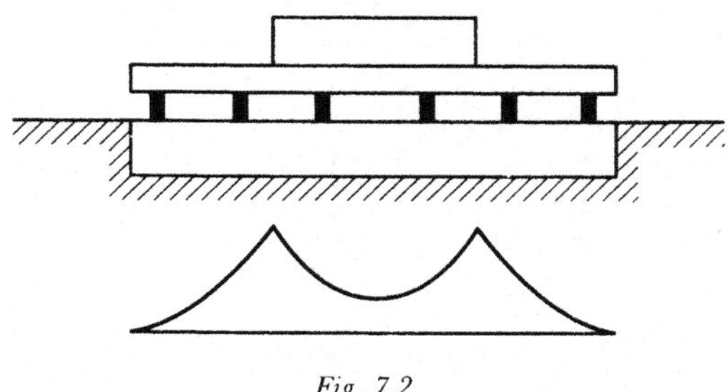

Fig. 7.2

can be plotted. Whatever the mounting arrangement some transfer of vibrations is inevitable and a plot such as Fig. 7.3 shows the importance of avoiding operations around the resonant state.

Transmitting amplitude is less than forcing amplitude for frequency ratios above $\sqrt{2}$ and for precision work an arrangement equivalent to a spring isolated system is required.

Joints should receive particular consideration in terms of load distribution in bolts, although an assembly consists of a number of joints in series the overall stiffness can reasonably be related to a solid bolt because of the pre-loading and high stressing which occurs in practice.

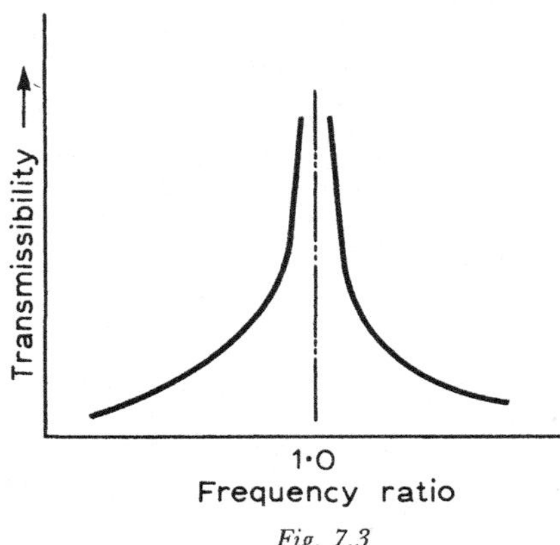

Fig. 7.3

7.4. MODEL ANALYSIS

Machine tool elements can be simulated by a combination of basic
beam and column shapes so that the behaviour of a prototype
structure can be predicted from model analysis provided that certain
relationships are known. Dimensional analysis can be used to
interpret test results, variables should be arranged in dimensionless
groups and the performance of one group analysed while other
group values remain constant. Model material is chosen in terms of
the result required, typically a stress or deformation situation. Geometric similarity is arranged by a scale based on the length relationship

$$C = L_P/L_M$$

where P is prototype and M is model. Alternative relationships might
be expressed for scalar values of load or oscillation. For a model
beam in static bending see Fig. 7.4 area

$$A_M = B_M D_M \text{ so } A_P = CB_M CD_M = C^2.A_M$$

and similarly for second moment of area

$$I_P = C^4, I_M$$

For elastic relationship in bending

$$d^2y/dx^2 = M/EI \text{ and load ratio } X = W_P/W_M$$
$$(\text{second differential})_P = (1/C)(\text{second differential})_M$$
$$M_P = L_P W_P = CL_M XW_M = CXM_M$$

and substituting for second differential as

$$M/EI \text{ gives } C^2/X = E_M/E_P$$

When an original design has to be up-graded to carry a higher
load or when fabricated methods are to be investigated the concept
of equivalent sections should be utilised. Copying of shape and style

is a common mistake as the form design of a component made from steel must be based on principles quite different to those used for iron castings. The basic steps are:

1. Determine strength and rigidity requirements.
2. Determine properties such as section modulus of cast member.
3. Use nomogram or section tables to obtain properties of steel member, a typical form is shown in Fig. 7.5.

A typical result is that the section modulus for structural steel in

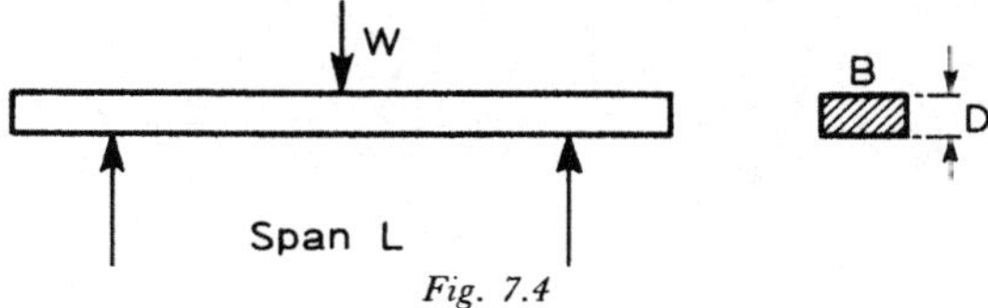

Fig. 7.4

a bending situation need be only 21% of that for low-grade cast iron.

An analysis of bending and torsion situations in the design of weldments typifies problems found in the design of machine bases. A force applied transversely to the axis of a partially supported member sets up bending moments along its length, and a horizontal shear stress arises due to variation of bending moment.

The most efficient use of steel in bending is obtained by placing flange material as far as possible from neutral axis, by welding ends to supports and by placing joints in low stress areas. Fig. 7.6 shows the crucial effect of section depth; notice that thickness is reduced

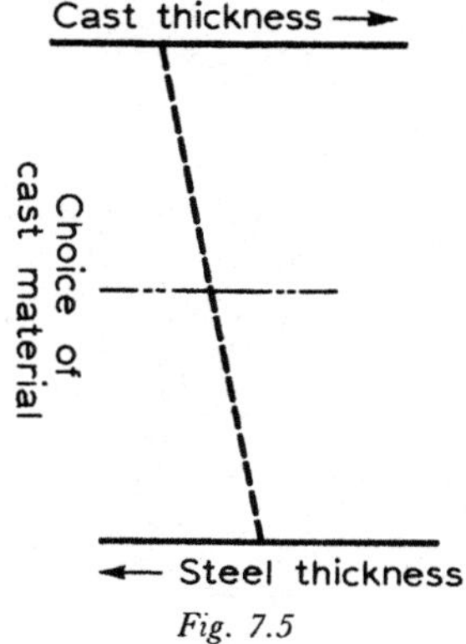

Fig. 7.5

to the extent that flanged sections soon become necessary. The basic rule for the best use of steel in torsion is to provide closed sections to carry transverse and longitudinal shear stresses, but careful attention must be paid to the positioning of braces and stiffeners particularly adjacent to access holes or apertures.

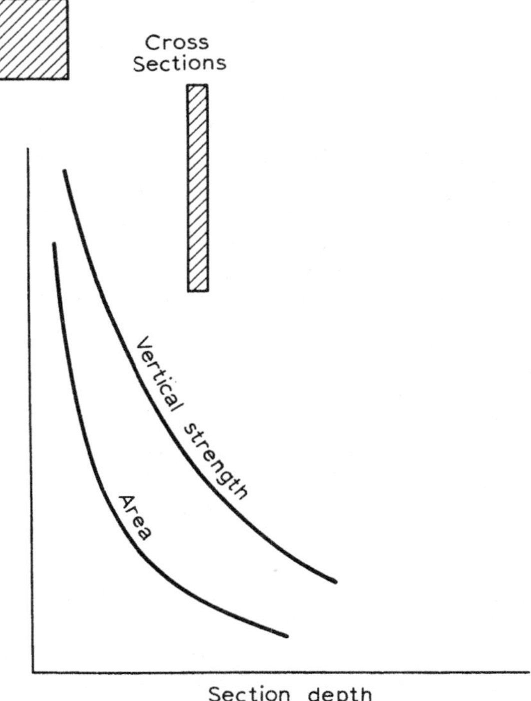

Fig. 7.6

7.5. BEARING CONSIDERATIONS

A number of interesting developments have taken place in the
provision of hydrostatic bearings for machine tools. A particular

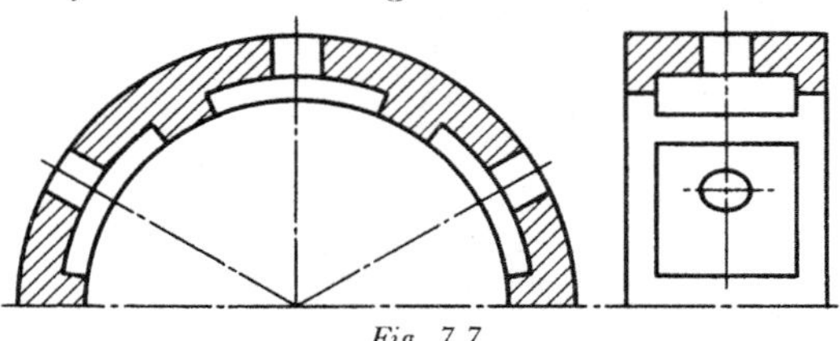

Fig. 7.7

requirement for spindles is the maintenance of rotational axis within
close limits in spite of load variations and thermal effects. Operation
of a bearing of the type shown in Fig. 7.7 is due to pressurised oil
from a pump flowing through restrictors into pockets so as to maintain a permanent film between the surfaces.

Use of restrictors reduces effect of load changes on film shape and radial clearance may be much larger than in rolling or hydro-dynamic types. Thus, a reduction in viscosity (for hydrostatic types) does not produce a significant reduction in radial stiffness which is such a problem in the other types.

In a sliding hydrostatic bearing of the type shown in Fig. 7.8 the supply is fed to a shallow recess or source, the intention is to provide

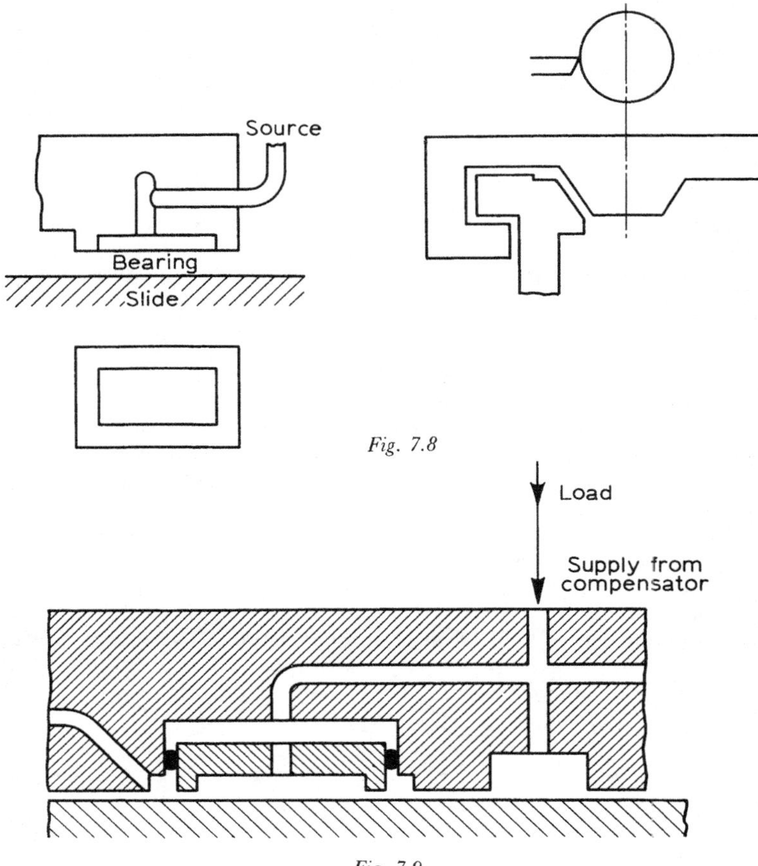

Fig. 7.8

Fig. 7.9

a laminar flow so that under static conditions the pressure/flow rate ratio should depend only on clearance and viscosity. Provided that supply is controlled with respect to bearing pressure the static per-formance may be predicted from a clearance or stiffness/load curve. It is necessary to distinguish between yielding and rigid systems; in the former the structure is permitted to deflect an amount which will not cause displacement at cutter to exceed acceptable limits. In a so called rigid system the bearing clearance is kept constant so that the supply pressure must provide a flow in proportion to bearing pressure variations.

An alternative arrangement is shown in Fig. 7.9 and incorporates a series of pressure chambers spaced to resist tilting in addition to normal load, each of these is controlled by a compensating cylinder which alters pressure in accordance with load variation. The size of chamber, compensator and supply pressure might be chosen to provide a clearance of 0·001 in (0·00025 mm) at maximum load.

EXERCISES

1. Classify the various orders of generation of a machined surface in terms of the linkage between workpiece and cutting tool.
2. Compare in kinematic terms the use of flat and vee bed guiding elements in metal cutting machines.
3. Discuss the three basic requirements of a machine tool and list operational features which have to be considered.
4. Establish equations for optimum cutting speed during rough turning on a centre lathe.
5. For a given stress and permitted deflection in various materials show that an optimum structure ratio of $\text{length}^2/\text{depth}$ can be determined.
6. Discuss the effect of bolted joints in machine tool structures in terms of their influence on stiffness and deflection.
7. Distinguish between the three types of vibration found in metal cutting machines and localise the parameters involved in the assessment of chatter. Sketch a typical stability chart for a regenerative chatter situation.
8. List parameters to be considered in the design of a hydrostatic bearing system with particular reference to the advantages expected in relation to machine tool applications.

8

TREATMENTS OF FERROUS METALS

The earlier chapters of this book have dealt with fundamental manufacturing processes in terms such as to permit a choice between processes to satisfy specific requirements. At this stage it is appropriate to summarise properties of the more common ferrous materials, either as raw material or ready for use, performance ranges on which design decisions can be based and the effect of various treatments to fit such materials for particular duties.

8.1. CAST IRON

Mechanical properties of cast irons depend upon chemical composition, rate of cooling and the extent of formation of voids due to graphite. Application of load produces a complex stress system with elastic and plastic deformation of the matrix and strain at graphite cavities. Rate of cooling and section thickness determine graphite form, the product of a single melt might vary from a hard unmachinable iron to a weak, open grained structure depending on the relationship between the above parameters. In any assessment of load carrying ability it is important to relate section thickness to test piece diameters, on which strength values are based. BS. 1452 quotes grades $10 \rightarrow 26$ for grey iron indicative of tensile strength tonf/in^2, although such values do not exemplify fitness for particular applications.

The variation of Carbon Equivalent Value, $= \text{Carbon}\% + \frac{1}{3}$ ($\%\text{Silicon} + \%\text{Phosphorous}$) shown in Fig. 8.1 provides a useful correlation between composition and properties.

Designs in grey iron rarely allow stresses exceeding 25% of the tensile strength in a tensile situation but up to three times such values in predominantly compressive situations. Applications demanding a high damping capacity or subject to thermal stressing are best covered by low-modulus irons. Malleable cast irons are produced so that the graphite is precipitated as nodules with much

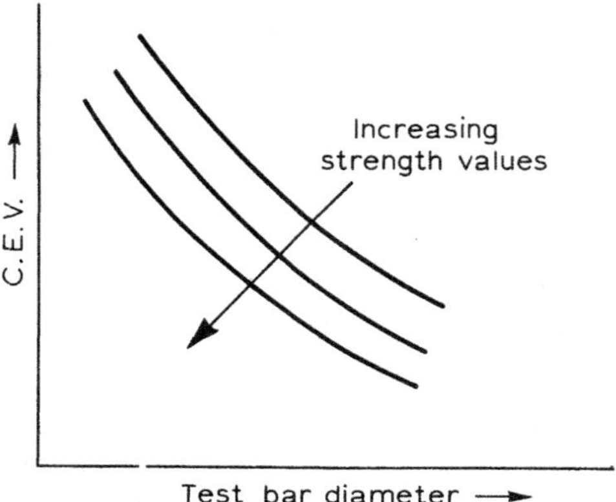

Fig. 8.1

less matrix discontinuity, a portion of their stress/strain curve has a linear relationship similar to steels. Such materials respond to heat treatment and can be hardened to provide tensile performance equivalent to medium carbon steels.

Comparison of specimens in terms of % pearlite is a convenient measure of strength and the rise from 53% in grey iron to 92% in proprietary malleable irons is indicative of performance improvement. The 0·1% proof stress is used to express properties, this value may approach the order of 60% tensile strength.

Much has been written as to the significance of hardness in irons, standard specifications list a considerable range of availability, values increase with tensile strength but not in a regular manner. The effect of this on machinability and wear resistance is obvious, but attempts to relate permitted stress to a measure of indentation have proved unreliable. Failure due to surface fatigue is found under large local stress so that load resistance has an apparent relation with hardness, few designers however would risk the theoretical premise of permitted stress increasing as the square of an indentation factor, only proportionality is considered safe.

In many real cases, demands on the material are based on considerations other than strength as the machinability concept often requires a section thickness which reduces stress to a very small value. The inherent presence of graphite results in a low notch sensitivity due to the many situations of stress concentration present in the structure, but careful correlation of values with the notched fatigue strength makes possible the specification of irons for crankshafts, which otherwise would require forged steel of much higher tensile strength. The subjection of grey iron to heat treatment has the dual purpose of relieving residual stresses due to the casting process and providing surface conditions suitable for subsequent machining. The range of treatment must be adjusted to maintain strength and to prevent the formation of an undesirable structure. For malleable irons the process is intended to increase strength as in steels.

8.2. STEEL CASTINGS

A particular feature which attracts the designer is the reduction in weight afforded by the use of steel castings, but section thicknesses must not be reduced to the extent of making casting difficult. Chapter 4 dealt with the casting problem and showed the dependence of thickness on surface dimensions, arrangements for gating and risering, etc. A minimum of $\frac{1}{4}$ in thickness for conventional steel casting is a useful guide providing that proper joint proportions and arrangements for directional solidification have been organised.

Steel is conventionally defined as an iron/carbon alloy containing less than 1.8% carbon, an alternative classification for carbon steels suggests those containing less than $1\frac{1}{2}\%$ manganese. The carbon content of steels commonly varies from 0.15% to 0.45%, values below the middle of this range are most easily cast but have a restricted strength range, hence the need for alloying to permit a desirable combination of strength and ductility. Cast and wrought steels of similar composition and heat treatment show little difference in properties such as tensile strength and hardenability, although directional properties will differ with the type and degree of working.

Alloying is frequently resorted to, chromium, nickel, manganese and molybdenum are common additions and may be classified in terms of their tendency to form carbide or to graphitise or to stabilise compounds. Their major function is to assist in producing fully-hardened sections due to heating above the critical range and cooling rapidly. Thus, the hardening process for ferrous materials is essentially a suppression of normal equilibrium changes with the intention of causing a martensitic transformation. The effect on the critical range may be represented in terms of phase temperature and cooling velocity or as in Fig. 8.2 when the strength enhancing capability of particular elements in a given material is represented.

The maximum hardness in a metal depends upon carbon content, various alloys will increase depth hardening ability, but maximum hardness will not exceed that of a steel having the same carbon content. Thus the heat treatment of steel, apart from structure

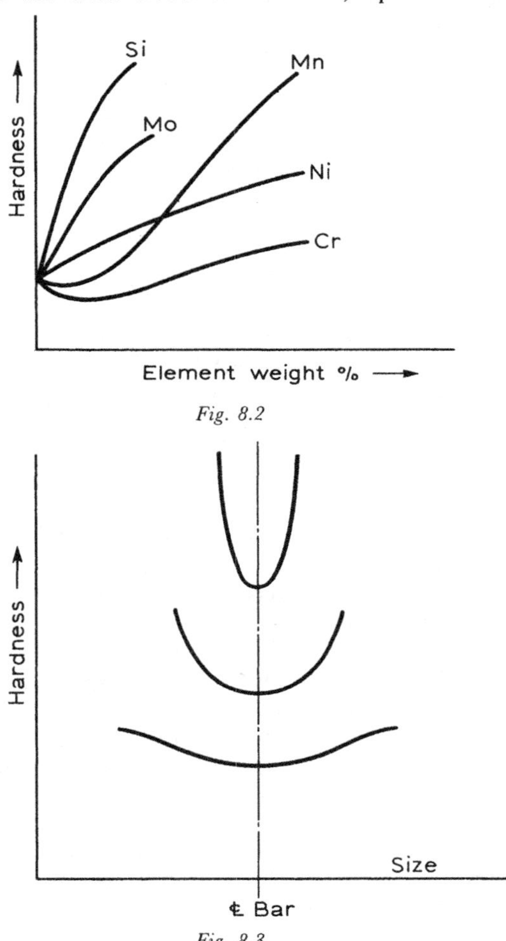

Fig. 8.2

Fig. 8.3

correction and machinability aspects, is mainly concerned with conferring improved properties. Alloying has just been referred to and the many surface treatments to provide skin hardness while retaining a sufficiently tough core will be considered later. Hardenability is a simple measure of the depth to which a desired hardness can be provided, or metallurgically to the likelihood of martensite being formed.

It is, therefore, a function of cooling rate, and BS. 970 and 971 express this property in terms of a ruling section. Notice the size effect in Fig. 8.3 the large variation in surface to centre hardness in a small bar is compared with the reduced fluctuation as bar size increases.

Another method of expressing the effect of quenching at various temperatures is by reference to an isothermal transformation diagram. Fig. 8.4 is typical of these time-temperature-transformations

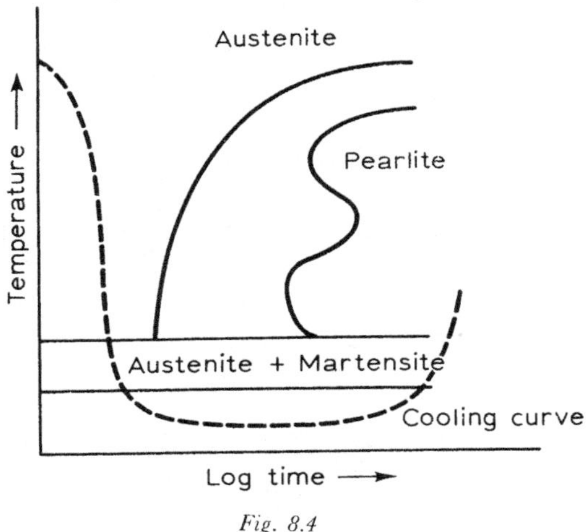

Fig. 8.4

or TTT curves and indicates that a martensitic structure, a super-saturated solution of carbon in alpha iron, an essential ingredient of hardened steel, is only obtained when quenching is sufficiently rapid to prevent the cooling curve intersecting the transformation curve. A typical test result comparing the cooling of a specimen in the furnace—where it produces 40% ferrite, 60% coarse pearlite—might show that 100% martensite is produced when water cooling is used.

8.3. HEAT TREATMENT PROCESSES

The remedial processes of normalising and annealing play their part in rectification of surface and structure, often as an interstage of a forming process and their effect on properties is shown in Fig. 8.5.

The hardening process should provide as high a value as possible consistent with the prevention of cracking and distortion in quenching; the cooling rate, which varies not only with composition but also with the quenching media, will determine properties and thus some knowledge of the heating and cooling change points is required. The range of availability is clearly seen on the iron-carbon

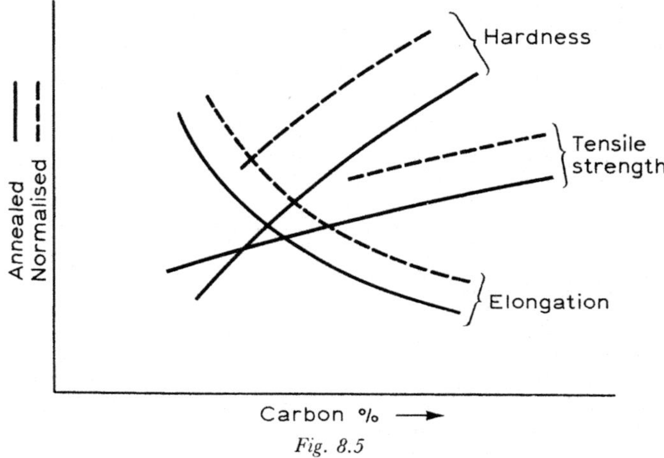

Fig. 8.5

diagram, shown earlier as Fig. 2.5, while Fig. 8.6 illustrates typical quenching behaviour. As the hot material is plunged it becomes surrounded by a vapour blanket and it is some time before liquid cooling commences.

The tempering process, over a much lower temperature range,

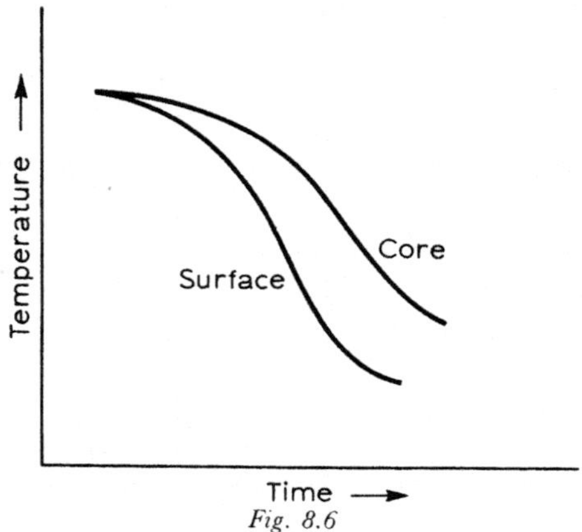

Fig. 8.6

has as its aim the retention of much of the strength indicative of a hardened profile, but provided as a property of a well-balanced structure, Fig. 8.7 shows typical property variations. The most common heat treatment error is the failure to provide sufficient soaking for solutions to develop, 1 hr/in (0·39 hr/mm) cross section in

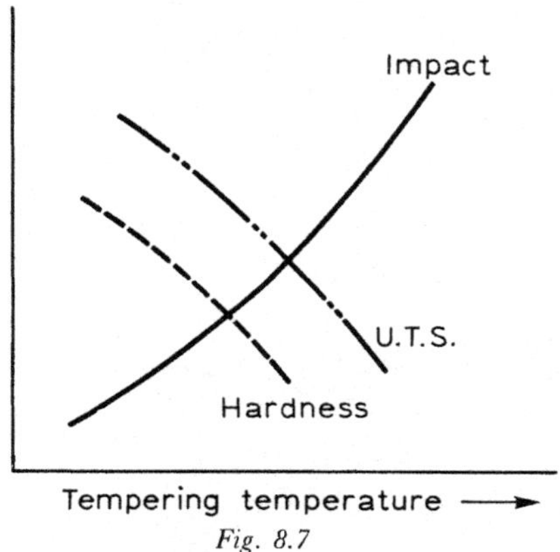

Fig. 8.7

hardening and up to 2 hr/in (0·78 hr/mm) in tempering are reasonable for components of large bulk.

From an economic point of view the selection process for a bar material should include the question 'will mild steel be satisfactory'? Its stiffness compares well with far stronger materials, plain carbon steels (0·12% to 0·25% carbon) satisfy many engineering specifications and might be regarded as the basis of many materials with considerably improved properties. BS. 15 (structural), BS. 970 and 971 (EN3 wrought), BS. 3100 (castings), are well-known versions quite apart from the ENI varieties supplied as free machining material. Variation in composition and treatment can provide ultimate stress ranges between $355 \rightarrow 494$ MN/in^2 and for some applications the saving due to easier machining represents a greater benefit than a cheaper material. Cold working and local carburising can provide an extensive variation of properties in mild steels.

Fig. 8.8 shows comparison of cost and ductility with a fully-hardened alloy steel, where HT is hardened and tempered, BD is bright drawn, C is carburised. When recourse has to be made to more sophisticated materials the golden rule is that strength and ductility are alternatives.

Case hardening or carburising is a well-proven method of providing a hard surface due to the entry of carbon to form a solid solution with iron. It is fortunately best suited to low carbon steels,

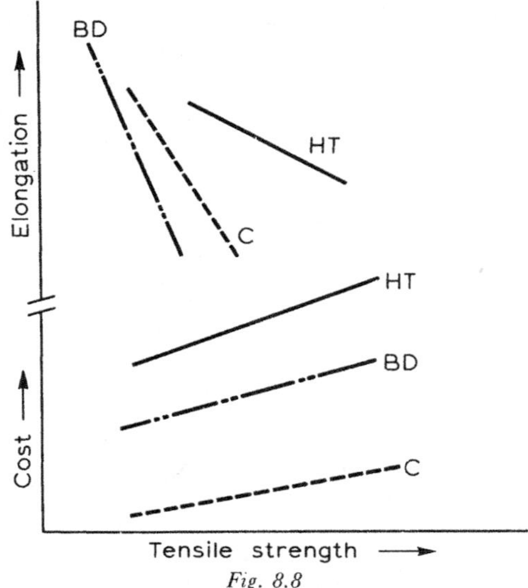

Fig. 8.8

whose surface properties are often deficient and the process can induce a skin containing 1% carbon, indicative of a hard wearing surface. Other common methods include:

Cyaniding which produces cases up to 0·020 in (0·51 mm) thick

Nitriding provides the hardest surface but not much deeper than 0·005 in (0·127 mm)

Induction and *Flame Hardening* is obtained by rapid heating and quenching, may require only a few seconds to penetrate to a depth of $\frac{1}{8}$ in (3·17 mm). The intention is to leave analysis of surface unchanged.

Carburising time is governed by the case depth required, a high power rating has to be provided to ensure short cycle time with high load density. A particular problem with rapid heating is lack of uniformity due to temperature gradients.

8.4. COST COMPARISON

It should not be thought that mild steel is necessarily the cheapest material, particularly when the machining programme is small, so that material cost and a relatively high strength would then be the criteria. On the basis of equal strength in tension, the weight of a 60-ton steel may be one-third and cost may be two-thirds of that for conventional mild steel. However, when stiffness is considered the 60-ton steel may be twice as expensive as the mild variety. Application of value analysis concepts are used to determine the cheapest method of producing an article while maintaining a minimum standard of quality. The definition of this technique for identifying and eliminating unnecessary costs is a fitting end to our consideration of ferrous materials, the aim has been to emphasize at all times the essential relationship between requirements and the most economical process method available.

EXERCISES

1. Compare the effects of heat treatment processes on the structure and properties of plain and alloy carbon steels. Classify the alloying elements in terms of their tendency to form or graphitise carbide.
2. Construct a table classifying eutectoid steel structure in terms of cooling rate and show how the results of Isothermal Transformation diagrams can be utilised.
3. List factors which limit the accuracy of hardness tests and summarise proposals to establish a relationship between indentation hardness and tensile strength.
4. Explain the need for heat treatment of grey iron castings and the results expected from such treatment.
5. Discuss the effect of size and shape on the depth of hardening of steel components, what is the significance of surface to mass ratio?
6. Discuss diffusion in terms of the temperature dependent migration of atoms and the role which this plays in annealing.
7. Consider the effect of prior cold work on the grain size and structure resulting from an annealing process.
8. Describe one process of case hardening based on carburising, summarising the heat treatment process which will be required.

9

NON-FERROUS CLASSIFICATIONS–LIGHT ALLOYS

Continuing the general consideration of material and process selection it is time now to refer to the specification and properties of non-ferrous metals. Unfortunately their classification is not so simple as for the ferrous types already discussed as there is no equivalent of BS. 970 and 971 for non-ferrous metals and their documentation often demands prior knowledge of trade names or extensive practice in the use of a standards year book. BS. 1400 is a typical example, providing specifications for ingots and castings covering thirty-three alloys. Much useful data can be found in handbooks of the various material development associations.

There are many anomolies, the name *aluminium bronze* is given to a material which contains no tin, *nickel bronze* and *nickel silver* are really a gunmetal and a brass respectively, *manganese bronze* is a high tensile brass. Nevertheless some indication of scope can be given by grouping in terms of the common base metals, not necessarily in terms of the forms in which they are available, since many alloys would appear in several categories.

It is perhaps in the non ferrous field that a closer study of materials data systems will pay dividends, but the scheme must be comprehensive since many materials are still based on aircraft and defence specifications, upwards of 2000 entries may be necessary for full documentation. The considerable influence of form and condition on properties is most marked in this context, ruling section is a vital parameter when properties vary over such a wide range.

It is of interest to compare the cost of ferrous materials, on a volumetric basis, with certain non-ferrous types discussed in this chapter. The table below is based on magnesium alloy ingots as *unity*, although not all the materials listed would necessarily be available in ingot form. Slight fluctuations in these ratios will occur depending on form and treatment.

Aluminium alloy	1·15	Copper alloy	10·5
Zinc alloy	2·3	Phosphor bronze	10·5
Manganese alloy	3·7	Nickel	14·2
Brass	5·3	Tin	22·6
Aluminium bronze	6·8	Beryllium copper	36·3
Gunmetal	9·2		

Contrast these values with a relative cost scale factor of about 1·15 for mild steel and it will be clear that the economics of selecting a non ferrous material starts off with a price disadvantage. Thus size, weight and environmental parameters must provide adequate reasons for any change from ferrous materials.

9.1. PROCESS PARAMETERS

Mechanical processing of non-ferrous metals and alloys differs little from the methods already described, except that they are of smaller scale and the greater ductility demands less power. There is scope for a wider range of forming operations than would be economic with ferrous metals and a considerable range of finishing processes since many non-ferrous products are finished bright. A desired finish on composite materials, which would be too weak or expen-

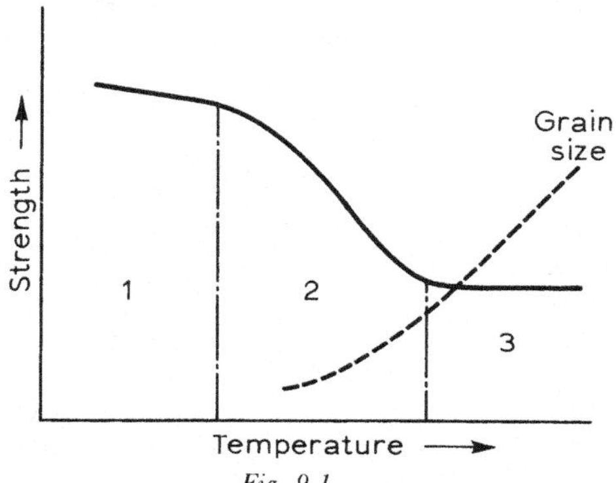

Fig. 9.1

sive if entirely non ferrous, can be provided by cladding or similar processes.

In some circumstances alloying will secure required properties without heat treatment but, in general, process treatments must be more precise. Knowledge of the relationships between heat treatment and cold working processes is the key to a basic understanding of the performance of non ferrous alloys. Well-marked phases can be defined such as (1) stress relief, (2) recrystallisation, (3) grain growth, as shown in Fig. 9.1. Brittleness is sometimes a characteristic of the cast state and annealing for homogeneity by diffusion may

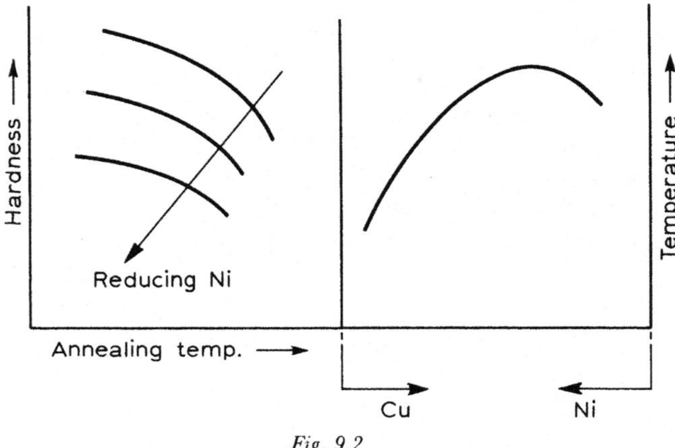

Fig. 9.2

be necessary before a deformation process can take place.

A convenient plot for particular material combinations relates annealing characteristics to the forming range. Fig. 9.2 for a copper/ nickel alloy shows how a temperature might be specified to restore a particular condition after cold working.

9.2. DIE CASTING

The extensive use of *die casting* for light alloys deserves particular mention here. A regulated pressure is vital for injection of molten metal into the die when thin sections are to be cast, flow must be regulated to prevent appreciable chilling and pressure maintained during solidification to help in reducing shrinkage. In the hot chamber, low-pressure process, the injection cylinder is immersed in molten metal which rules out any material whose reaction may contaminate the melt, this is used for low melting point alloys of zinc, tin and lead. The former is the principal base metal in order of commercial die casting exploitation, most competitive for parts weighing up to 133 N, and accounts for probably 70% of the total.

For alloys of higher melting point; brass, aluminium and magnesium, an auxiliary heating method is necessary. Considerable pressure must be provided to force metal through a nozzle into the die and heat-resisting alloy steels must be used for die construction. Any decision to use the die-casting process must be largely economic; dies are expensive although the better quality materials might produce 600 000 pieces in zinc and there are restrictions on shape since die moulds cannot be disposed of as in sand casting. Thus, optimum batch numbers of 100 000 are not unusual although the particular process advantages of a smooth, accurate finish, considerable reduction in subsequent machining requirements and uniformity of structure mean that smaller batches may be regarded as viable.

Structure uniformity is not achieved by accident and the nature of the process demands careful attention to die design and feeding arrangements. Since cooling progresses from die walls towards centre, the ensuing contraction means that replacement is from a hotter source, which tends to cause a void or shrinkage porosity. The extrusion process finds its most successful usage in the field of metals having large ductility and plasticity. There is extensive use of this process for non-ferrous alloys particularly in the range of shapes unsuitable for rolling.

9.3. BASIC ALLOYS OF COPPER

Copper and its alloys form the largest numerical section of non-ferrous metals, twenty-three standards are currently needed to specify the more common basic forms and alloys. Base metal properties, as cast, might provide tensile strength 154·4 MN/m^2, elongation 25% but a moderate cold work reduction would double this strength and halve elongation, and the exceptional plasticity is well documented in respect of hot working operations.

The most common alloys with copper are zinc, aluminium, tin, nickel and the behaviour of a crystalline lattice will be considered in respect of the first of these. The original lattice is f.c.c. (unit cell 4 atoms), and the addition of a zinc atom in solid solution produces an α solid solution which does not essentially alter the lattice. The distortion due to the introduction of a second element will harden and strengthen at the expense of ductility, similar to the situation resulting from lattice distortion due to deformation. Fig. 9.3 shows a portion of the copper/zinc equilibrium diagram and illustrates the variation of properties.

Two main groups of copper and zinc are the basis of the industrial range of *brasses*. Fig. 9.3 shows also the relationship between the α and β groups. The α series are relatively ductile and with careful annealing considerable cold work is possible. The β series are

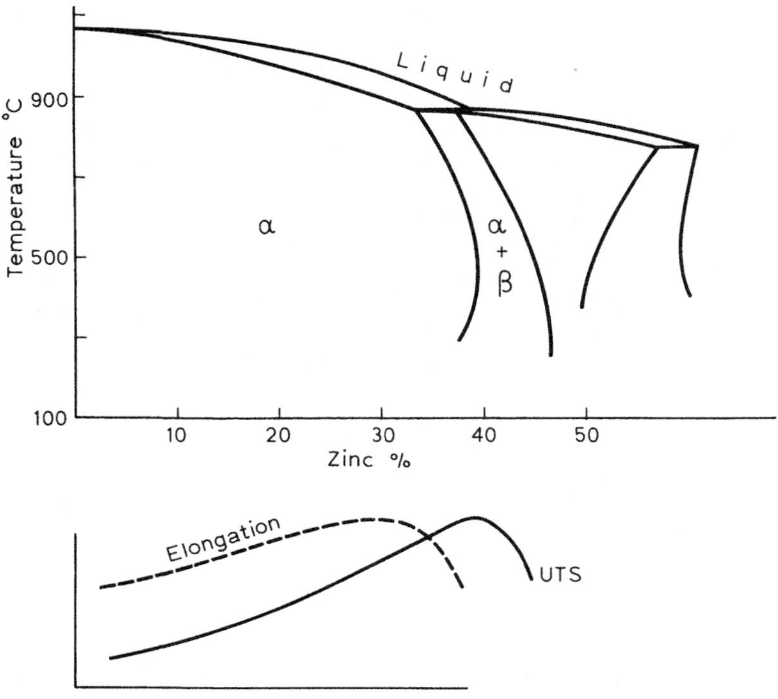

Fig. 9.3

nominally hotworked because of their sensitivity to cold deformation and often form the basis of other alloys with aluminium, iron or manganese.

There is wide usage of $\alpha + \beta$ brasses, 60:40 is a typical composition providing high tensile properties and manganese bronze is one example. Other special brasses with inclusions of nickel and aluminium can be temper hardened to provide specific properties. True *bronzes* are an alloy of copper and tin but the family includes metals with the addition of zinc (and perhaps lead) to form *gunmetals* and phosphorous to form *phosphor bronze*. When both zinc and tin are present, the greatest composition of one of these is sometimes used to define as a brass or a bronze. A considerable improvement in conventional bronze bearing properties results when lead is added at the expense of copper and tin, although strength is much reduced and the alloy may be used as a facing on a stronger back.

Fig. 9.4 shows the complex nature of the bronze equilibrium diagram and the many phases which occur in the wide separation between the liquid and solid states.

A range of so called bronze alloys based on copper and aluminium often contain nickel and manganese and are characterised by corrosion resistance and excellent resistance to elevated temperatures. Cold working is usual for strengthening purposes, although many materials can be formed hot and considerable modification

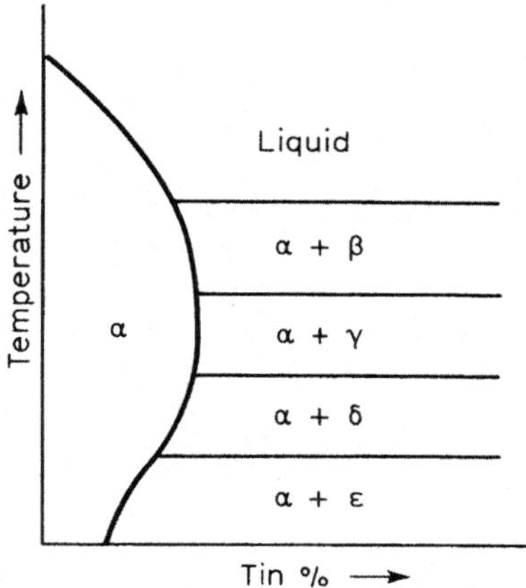

Fig. 9.4

of properties is possible by controlled heat treatment. Small additions of *beryllium* to copper produce a hard alloy, responsive to heat treatment and by appropriate relationships between such treatment and cold work a performance equal to that of ferrous alloys is obtained. Copper and nickel mixtures provide a continuous

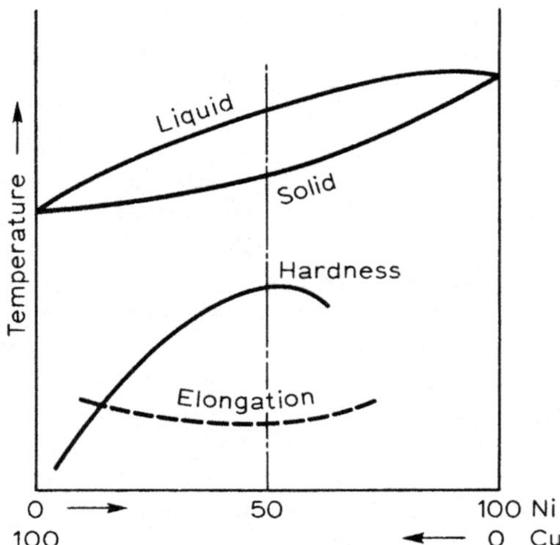

Fig. 9.5

series of solid solutions which are malleable in the cold state, proprietary metals with aluminium and silicon additions provide opportunities for precipitation treatment which greatly increases

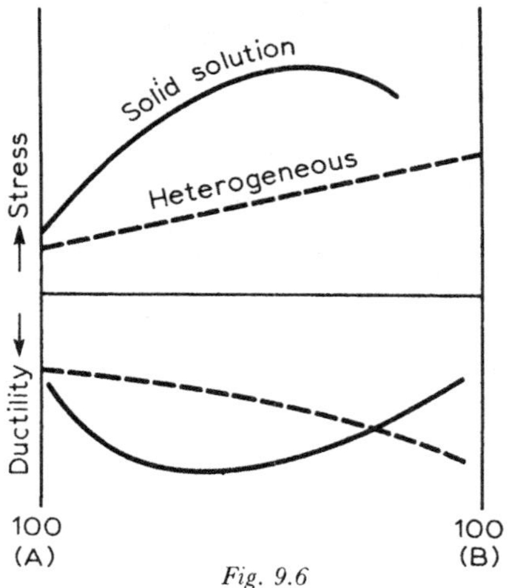

Fig. 9.6

hardness; retention of strength at high temperature is another feature.

The copper/nickel pairing is a good example of a binary alloy, formed from metals which are soluble in both liquid and solid states. Fig. 9.5 shows that it provides an entirely different type of equilibrium diagram, although in many of these mixtures the theoretical equilibrium is not realised due to rapid cooling. This results in a variation of concentration from centre to periphery of section, the core being richer in the highest melting point constituent.

Special attention should be paid to the cold forming properties of binary alloys. Fig. 9.6 shows the dependence on element properties and their variation with structure for two types of solution. The addition of up to 5% iron, manganese and silicon combined provides an opportunity to improve tensile properties while retaining ductility and the high temperature properties are excellent. Solution treatment is often referred to in the preparation of non-ferrous alloys. This involves heating along the line YY as shown in Fig. 9.7, so that an alloy originally composed of $\alpha + \beta$ solids has the β phase dissolved. Such diffusion is slow and its reverse involves precipitation out of solution which is even slower, variously called age hardening or precipitation hardening, depending upon the temperature range. In commercial processes

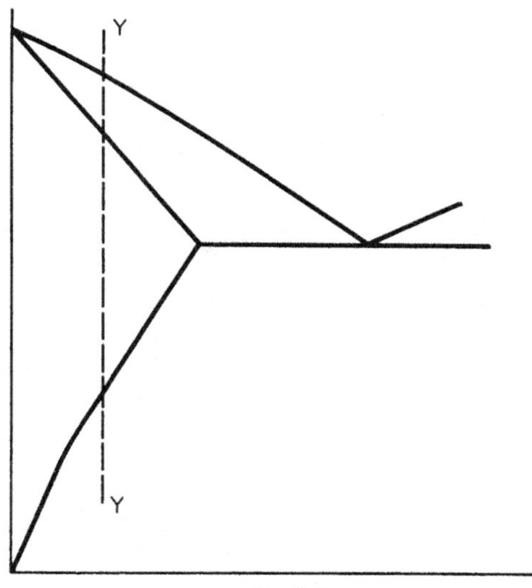

Fig. 9.7

solution treatment is followed by quenching along a phase line so that age hardening results from precipitation. The basic requirement for an alloy which will harden by precipitation is the decreasing solubility of one component in another with decreasing temperature.

9.4. LIGHT ALLOYS

The suitability of the light alloy range for prime consideration in any list of engineering materials has already been referred to, quite apart from weight saving and characteristics such as conductivity the corrosion resistance which many of these metals possess permits the designer to dispense with expensive coatings necessary for other materials. Pure aluminium and magnesium are not suitable for load carrying but there is a vast range of alloys, not only for strength improvement but also to improve forming characteristics. These are codified in BS. by letter and number symbols indicating the form and treatment, BS. 2970 and 3373 are typical for magnesium and BS. 1470 and 1490 for aluminium. Reference has been made earlier to the difficulty of establishing economic criteria for the choice between light alloys and steel. The specific strength concept employs the ratio strength/specific gravity and will show a 2/1 factor in favour of some light alloys, but in terms of equal stress such alloys may deform three times more than steel within their elastic range.

All aluminium alloys will harden in cold working so that anneal-ing services need to be provided in order to prevent a dangerous reduction in ductility, a range of $340 \rightarrow 400°C$ is typical. Uniformity of heating and precise knowledge as to recrystallisation temperature

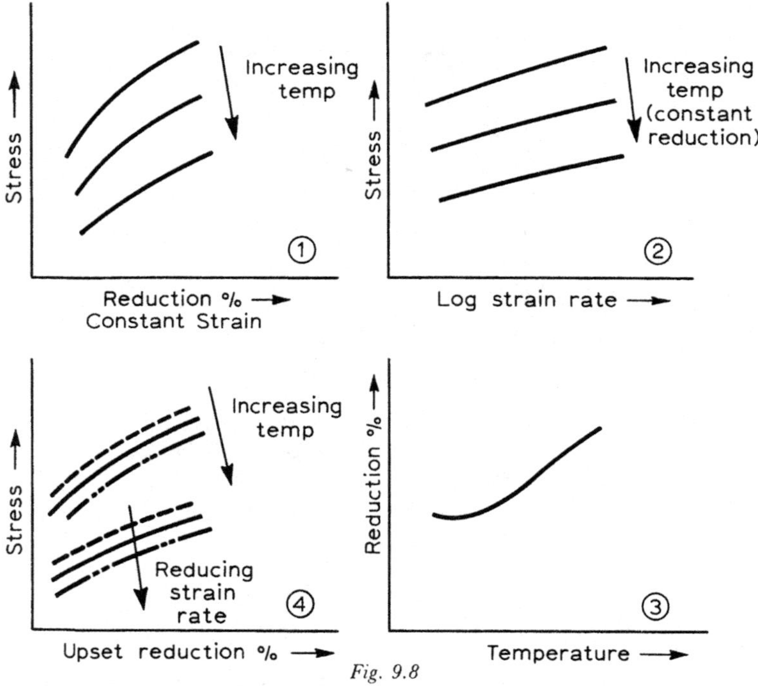

Fig. 9.8

are essential. Full heat treatment may involve both solution and precipitation treatment aimed at capturing the characteristic age hardening with its improvement in hardness and strength. Dur-alumin is a well known precipitation hardened alloy which takes up to four days to reach full hardness at room temperature. Aluminium alloys are readily forged to precise shapes, there are few restrictions other than those set by die design.

Fig. 9.8 shows a useful family of curves from which forging be-haviour may be predicted, for example section 4 leads to empirical relationships such as a forty-times increase in strain rate being roughly equivalent to a temperature reduction of say 93°C. Data on elongation at elevated temperatures may also serve as a guide to forging temperature range by identifying phases of recrystallis-ation and grain growth.

9.5. MAGNESIUM ALLOYS

Magnesium is the lightest metal in regular use, its density about two-thirds of aluminium, but with an elastic modulus about 20% of steel. It has a strength deficiency for most applications, alloying with aluminium and zinc is common for cast and extruded products

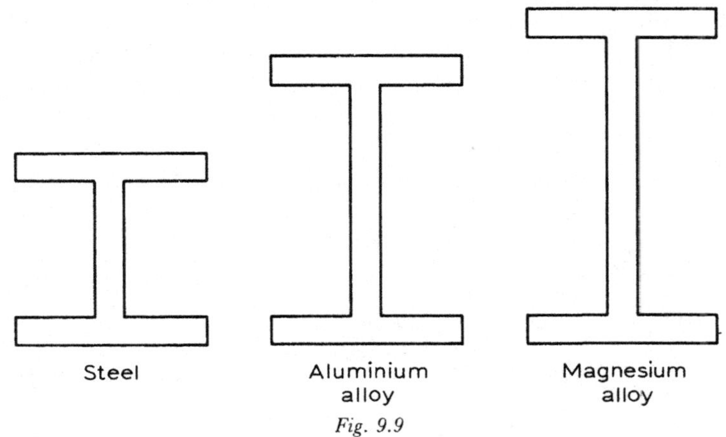

Fig. 9.9

or with zirconium for cast and wrought forms when a refined grain size is specified.

An interesting comparison between sections of equal width and stiffness is shown in Fig. 9.9 the magnesium alloy having a weight 37% of the steel section.

As is usual with light alloys the design process must allow for particular characteristics such as notch sensitivity. The temptation to try and solve rigidity problems by increasing wall thickness does not have the same success as with ferrous materials. Manufacturing methods such as die casting rule out thick sections and it is common to find wall thickness empirically related to surface area. In fact the production of magnesium alloy components encompasses most of the forming processes, ingots converted to billets by extrusion and subsequent bar rolling is frequently found although some forging processes are difficult with wrought alloys, slow pressing being preferred.

Forging exercises a key role in establishing uniformity of properties and ductility. Room temperature properties are strongly dependent on forging procedure and when rapid grain growth may be a problem each operation is forged at successively lower temperatures.

9.6. ZINC ALLOYS

Any large scale use of zinc alloys will involve consideration of die-casting principles, which have been discussed earlier, BS. 1004 lists composition of the common alloys. With these materials there is need to carefully control the presence of alloying elements which exert a marked influence on inter-crystalline behaviour. Dimensional and property changes due to ageing need sometimes to be taken into account, a shrinkage of 0·0006 in/in over 12 months might be associated with a 7% reduction in tensile strength but a 50% increase in elongation facility.

EXERCISES

1. Draw up proposals for a classification system for non-ferrous materials in terms of the common base metals.
2. Outline the outstanding characteristics of solutions and the distinguishing features of isomorphous alloy systems.
3. Compare common forming processes for ferrous and non-ferrous metals in terms of their scale or size, noting the differences in technique.
4. Explain why the relationship between heat treatment and cold working is so important for non ferrous alloys.
5. Compare the working properties of 70/30 and 60/40 brass, explaining the effect of small additions of lead to the latter.
6. Discuss why aluminium alloys can readily be forged to precise shapes and how mechanical properties may be developed by duplex heat treatment.
7. Explain why the room temperature properties of magnesium alloy forgings are dependent on the forging procedure.
8. List the important characteristics of a metal to be formed by die casting, noting any restrictions which this method imposes.

10

POLYMER FORMATIONS AND CLASSIFICATION OF PLASTICS

Before considering materials for specific transmission applications, reference must be made to the continued increase in the development and utilisation of plastic materials. We have certainly not seen the full scope of recent improvements in cost comparisons, while the limitations in strength capacity, which once formed a barrier to the extension of plastics usage, have been to some extent removed by developments in reinforcing and filling techniques. The title 'plastics' originating from the early polymers is no longer descriptive, many modern derivatives exhibit characteristics associated with hardness and brittleness. To some extent these developments have been brought about by the realisation that plastics are not merely a substitute for metals but have their specialised field in which they are superior and in other fields certainly competitive. Of course, the methods used for design involving metals are not suitable for plastics and a drastic rethinking is necessary before adaptations of conventional design theories can be employed.

In the interests of providing a grounding in plastics technology suitable for young engineers, one of the objects of this chapter is to provide a simple explanation of the background on which further studies can be based. An understanding of the formation of large molecules from smaller ones, i.e. polymerisation will be required, the simplest reaction known as a linear addition is the basis of the long chain type molecules associated with polymers.

10.1. ATOMIC BEHAVIOUR

Most people are familiar with the behaviour of materials whose small molecules take on an ordered arrangement in solid/liquid changes. These are referred to as crystalline solids and their melting temperature is easily predictable when sufficient thermal energy is available to overcome inter-crystalline forces. When the molecules are large the structure increases in complexity, the arrangement is less ordered and crystallinity is a relative term, being found only in small regions within a structure which is generally amorphous. In any event the amount of crystallinity in plastic materials is not a constant and is easily changed by alteration in the cooling rate.

An important influence of the rate of crystallisation is the ease with which a chain segment can rotate, the cessation of such movement is called the glass transition point (GTP). Below GTP the

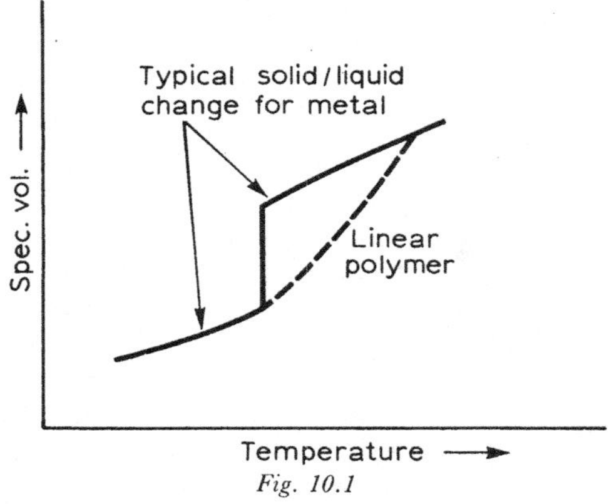

Fig. 10.1

material may have a polymer structure of short range order similar to glass, but above it the structure is similar to a viscous liquid or a rubbery material, it becomes tougher and more resilient.

Typical changes are shown in Fig. 10.1 and on passing through this phase, there is a marked reduction in elastic modulus. The key to the tremendous growth in plastics usage was the commercial development of synthetic polymers. The detailed procedures are beyond the scope of this book but only a knowledge of simple chemistry is required to appreciate the nature and structure of organic compounds and inter-molecular forces. In organic polymers, carbon is recognised in a four-bond tetrahedral valency state and in the straight chain polymer molecule each carbon atom forms two simple bonds with other atoms in the chain and two with part of another group. Remember that there is free rotation about a single carbon/carbon bond and that the system changes with rotation about the bonds. Other than these carbon arrangements the general heading of plastics will embrace the range of silicones (as fluids, resins or rubbers) where certain carbon atoms are replaced by silicones, and the field of glass resins where the carbon links are replaced by inorganic radicals. Fig. 10.2 shows a simplified version of the plastics family tree, this is by no means exhaustive since there

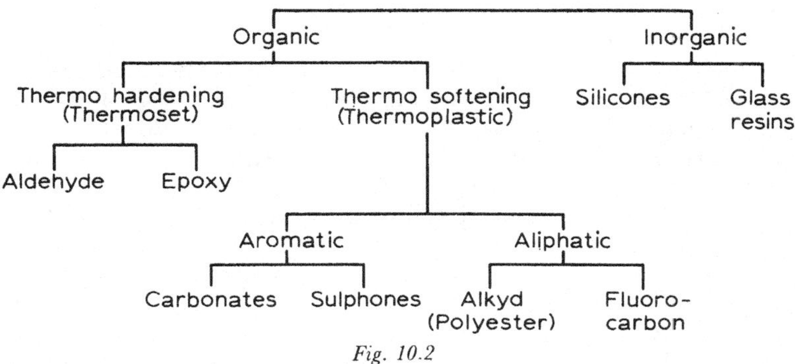

Fig. 10.2

are at least twenty-six groups of plastic materials in commercial development, and often a variety of grades within a group.

10.2. ADHESIVES AND COATINGS

Before dealing in greater detail with basic classification and properties the fundamentals of plastic adhesives and coatings will be considered. The basic ingredient of an organic coating is a mixture of polymers whose chief function is to provide protection and improve appearance. The vehicle or agent contains a vegetable or resinous binder which forms the film and holds particles of pigment, it often includes a volatile solvent or thinner to assist application. Thus, the major difference between paints is in the binder which controls the film durability and the mechanism of hardening, and it is the drying of binders which can be likened to polymerisation assisted by oxidation. The fundamental requirement for adhesion is intimate contact between surfaces, the agent must wet both surfaces to exclude air and voids. The material should be capable of application in liquid form and have a self-hardening facility. Semi-natural polymeric glues and varnishes are being replaced by PF thermosetting resins, either heat and pressure or room temperature, cured, while Polyvinyls (thermoplastic) adhesives are applied in solution or suspension to form a bond as the solvent dries. As an alternative to shellac and oil coatings cellulose derivatives, usually called lacquers, are widely used and condensation reactions between polyfunctional acids and alcohols provide coatings of the polyester family for the manufacture of modern varnishes and paints.

10.3. MATERIAL PROPERTIES

The relationship between processing methods and material properties is more critical in polymer applications than in those involving ferrous materials. It is necessary to guard against misuse of test data, particularly that based on conventional tensile and impact procedures, extrapolation methods which might succeed with metals are dangerous for plastics because of non-linearity between stress and strain. Some properties may vary appreciably over quite narrow temperature ranges and there is a noticeable increase of strain under constant stress, usually referred to as creep, which often prevents the definition of a point of yield. A convenient method of expressing this effect is to regard it as causing a gradual

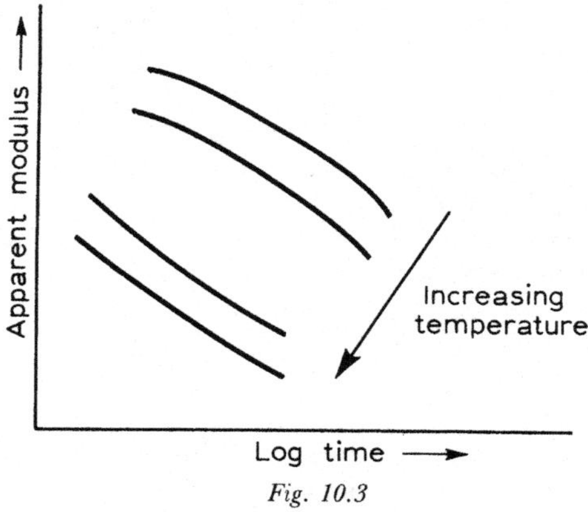

Fig. 10.3

reduction in elastic modulus, we thus refer to a time dependent, long term or apparent modulus which might vary as shown in Fig. 10.3. The time dependence of creep is also illustrated in Fig. 10.4. Notice the three typical stages and how removal of load leads to some elastic recovery, but not to the extent of preventing a permanent set.

For engineering applications few plastics with a short term modulus below 6.895×10^5 kN/m^2 are likely to be satisfactory and there may be design restrictions if the stress/strain curves show marked non-linearity for strain rates over 1%. Since long term test data is not available for many materials, reference is sometimes made to empirical ratios between long and short term moduli, 0·25–0·5 is typical. This emphasizes how unwise it is to visualise safe stress as a fixed proportion of short term test data, particularly when based on impact behaviour since notch effects are uncertain, thus the application of a large safety factor is only compounding the general

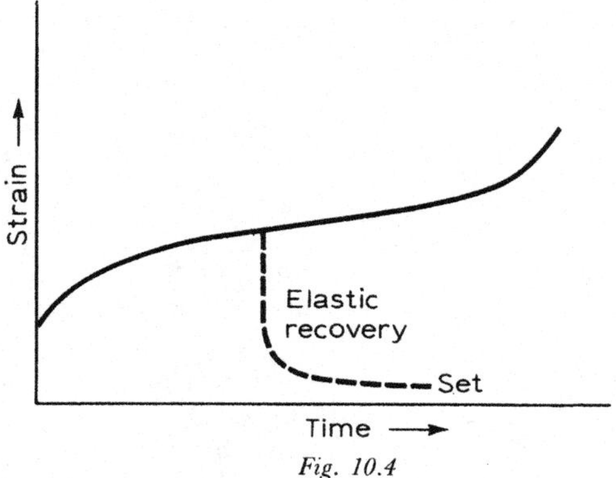

Fig. 10.4

ignorance as to safe working values. If, however, long term values are available a minimum safety factor of 3 should suffice. The relatively large deflections found when plastic materials are loaded, tends to negate the convenient assumption of small deflections in the application of the bending theory, there may be reason to invoke the large deflection theory for design purposes. This means that it is not advisable to substitute the theoretical flexural modulus in beam formulae since such values tend to be related to particular ambient conditions. Similarly any assumption as to working stress

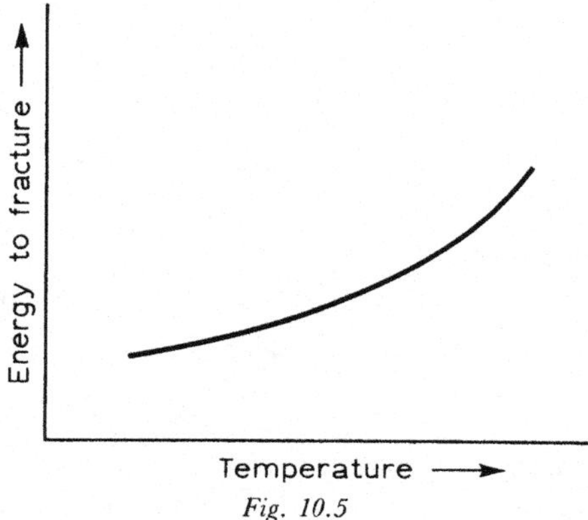

Fig. 10.5

values ought to be related to the care taken in manufacture to avoid residual stresses, or an undesirable orientation to anistropy—the change in physical properties with direction.

Conventional impact tests do not serve well for plastics, notch sensitivity is much more significant than for metals, so that performance results may have to be conditioned by shape and finish. Some authorities prefer to quote a low temperature brittleness point at some predetermined strain rate, Fig. 10.5 shows a typical relationship from which a safe working range might be determined.

10.4. THERMOPLASTICS

It is time now to refer again to the family tree, Fig. 10.2, and to consider in greater detail the two main divisions of organic polymers. *Thermo-softening* materials, usually called *thermoplastics* (TP) are characterised by softening on application of heat so as to reach a formable state before decomposition sets in, the formed shape being retained on cooling. There is no basic chemical change in the injection moulding process, (comparable to metal die casting) which is the usual method for production of TP and has replaced earlier casting techniques in which the resin was poured into a mould and cured *in situ* by the application of heat. In the injection process the charge is heated to a plastic state, transferred under

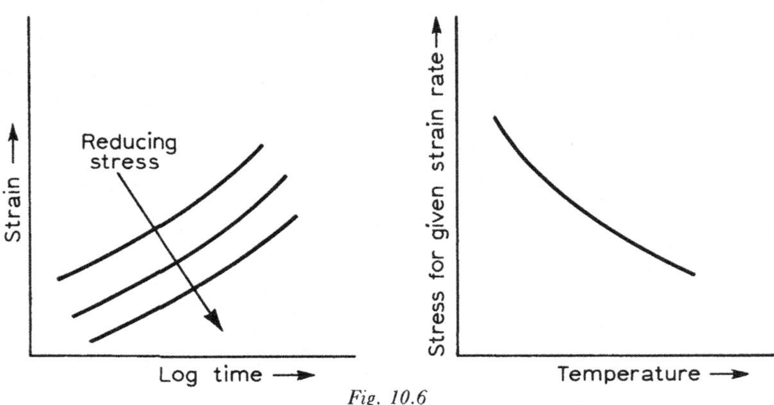

Fig. 10.6

pressure to a closed mould where it cools or cures before removal. The basic design requirement for shape is that a moulded piece must be extracted whole, rigidity is best obtained by judicious curvature and placement of ribs rather than from thick sections whose porosity may be in question. BS. 4042 deals with tolerances and allowances for particular processes. Extrusion techniques are widely used for rods and tubes, Blow moulding is convenient for bottles and hollow sections, thermo-forming or vacuum-forming is used for processing of sheets and films.

The effect of stress/strain behaviour with time and temperature is shown in Fig. 10.6 from which it is clear that designs based on Hookean behaviour are unlikely to be satisfactory. If the limiting

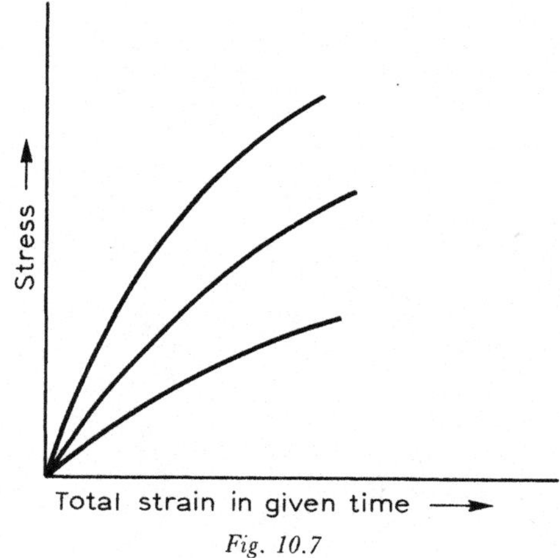

Fig. 10.7

strain for an application can be specified the allowable stress can be determined from isochronous curves of the type shown in Fig. 10.7.

A common definition for a TP is as a visco-elastic material, implying behaviour between the classical performance of a Newtonian Fluid and a Hookean solid. It is not the intention here to include a usage guide or glossary of thermoplastics materials in regular production. The list which follows is given so that the reader can recognise certain of the more common types. Some authorities prefer to write Poly outside the name bracket, such as in

Poly (Tetra Fluoro Ethylene) commonly called PTFE
Polyvinyl Chloride   PVC
Polyethylene   Polythene
Poly Methyl Methacrylate   Perspex
Cellulose Derivatives (Nitrate or Acetate)   Celluloid
Polyamide   Nylon

10.5. THERMOSETS

Thermo-hardening materials usually called *Thermosets* (TS) are characterised by a series of non reversible changes, they flow when heated but their shape cannot be altered after setting. Compression moulding is the usual (and certainly the cheapest) manufacturing process based on the application of pressure to fill a mould shape during the plastic phase. After heating of charge the pressure due to mould closure is maintained long enough to ensure curing the chemical cross-linking which occurs at a specified temperature.

The most common TS (Bakelite) is provided by the reaction of phenol (carbolic acid) and formaldehyde in the presence of a catalyst, the aldehyde resin or phenolic plastic which is produced tends to be amorphous both before and after polymerisation. A hard and brittle substance results from an unmodified reaction, which is of little use for load carrying members, so that elaborate filling or fibrous reinforcing techniques have to be applied to ensure desirable mechanical properties, including the use of metallic fillers for some cold curing resins.

A wide range of phenolic laminates is available, constructed from sheets impregnated with uncured resin in a suitable solvent, laminates are bonded under heat and pressure to provide a hard but machinable material. One range of glass-reinforced TS has a performance comparable to metals on a strength/weight ratio and are much less dependent than most plastics on temperature and rate of loading.

Epoxy resins, commonly called Araldite, are available as liquid or powder, frequently used as adhesives or for encapsulation, shrinkage during curing demands careful attention to filling techniques. BS. 2026 discusses the special problems associated with maintaining of tolerances, certain limits are imposed due to the problem of maintaining size during ageing.

10.6. PROCESS AND PROPERTY RELATIONSHIPS

There is, in theory, a very wide range of long chain polymers which might be used as plastics, but large scale production is confined to a small range of the simpler types we have discussed. Production of the four main thermoplastics amounted to some five million tons in U.K. and U.S.A in 1967, plus about 25% of this tonnage in thermosets. In trying to decide on a particular type for a given application the rigidity problem and dimensional instability associated with TP materials should not be overlooked, their high coefficient of expansion (probably ten times that for metals) requires careful attention, not only in terms of the working temperature range but also for assemblies involving other materials.

As a general guide for the machining of plastics a combination of high speed and small feed will produce the best results, the tendency to damage by overheating indicates the desirability of providing a coolant. The tendency to material instability causes flexibility of the specimen to be a problem, the additional support of box tools is often specified for turning operations. There may be problems in precision working if transition points produce dimension changes at temperatures within the machining range. Many laminated materials can be treated like non-ferrous metals although at the higher speeds possible it is unlikely that feeds exceeding 0·25 mm/rev will be used. Some plastics are abrasive and attempts to use conventional methods will result in uneconomic tool life.

Sufficient has been said to indicate the difficulty of predicting

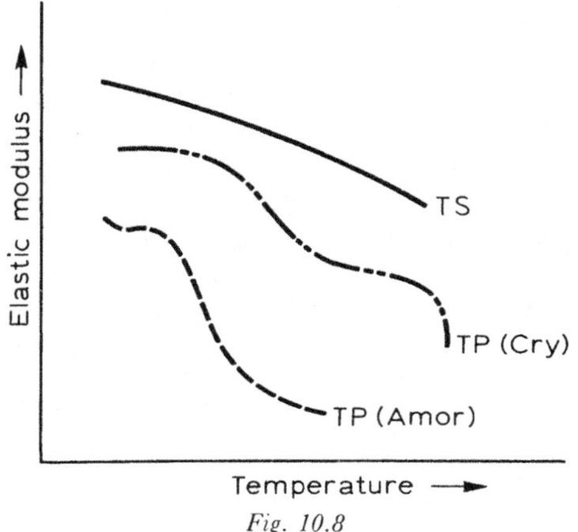

Fig. 10.8

properties from structure although generalisations of the type shown in Fig. 10.8 provide some guide.

Similarly the family of curves shown in Fig. 10.9 will reinforce the earlier remarks as to the importance of relating strength values to strain rate and temperature. Notice that a reduction in temperature may produce an effect similar to an increase in strain rate and that for all practical purposes a failure situation is reached at yield point in a plastic, there is no range of 'plasticity' common to many ferrous metals.

Some of the advantages of using plastics have already been referred to, the low density is some compensation for elastic modulus

(TS) probably less than one-fifth of that for steel so that a good strength/weight ratio is maintained. There are few problems in forming intricate shapes although the high cost of die and mould equipment demands long production runs. Thus a somewhat wider cost aspect involving tooling and maintenance has to be added to

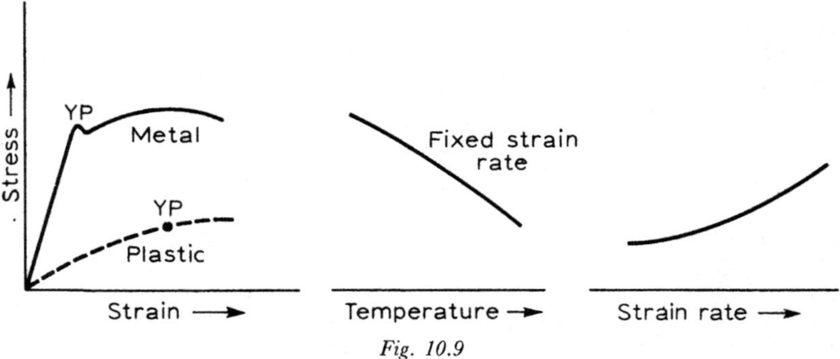

Fig. 10.9

the nominal cost of raw material and its conversion. A volumetric basis provides the most realistic comparison and over a recent twelve-year period the cost of a simple plastic (s.g. 1·4) reduced in price from 1·5 to 0·6 pence/in³ while steel ingots (s.g. 7·8) suffered a price increase from 7 to 10 pence/in³. Individual plastic costs vary widely depending on shape and number in batch, average cost pence/in³ might range from 0·6 for polystyrene though 0·8 for phenol formaldehyde to 25 for PTFC. Many of the more sophisticated plastics of engineering interest are too expensive except for special duties, thus the economic problem is a considerable one, frequently more difficult to resolve than in situations where ferrous materials are to be used.

10.7. TESTING CONSIDERATIONS

The testing of plastics is well documented in BS. 2782 which contains sections covering the effect of temperature on strength, deformation, flow and shrinkage, special tests for water content, flammability, density, etc., quite apart from conventional tests for electrical and mechanical properties. In view of the importance of the upper temperature limit on load bearing properties, particular attention should be paid to heat distortion tests (temperature at which deflection reaches a predetermined value), also to the softening point test (the amount of penetration by a loaded needle).

In view of the wide range of applications it is not surprising that the fatigue testing of thermoplastics has received particular attention. Polymers are prone to the development of internal stresses due to differential expansion with a temperature gradient within the specimen, accentuated by high testing speeds, which causes a higher stress in the skin for a given strain. There is also the problem, common to viscous materials suffering deformation, of applied energy being dissipated as heat instead of being stored as in a spring. Plastics are essentially poor conductors of heat, the hysteresis build up can profoundly affect properties as does also the sensitivity to dynamic loading, which because of the variable modulus makes prediction of a 'knee' on the stress/cycles curve most difficult.

10.8. RUBBERS

Finally, in any consideration of polymeric materials the various grades of natural or synthetic rubbers (or elastomers) must be included. $C_5 H_8$ isoprenes have a unique configuration, any change in which provides quite different properties. Natural rubber is Cis-Polyisoprene, an irregular structure with no crystallinity but containing centres of unsaturation suitable for addition polymerisation and cross linking. The commercial product contains fillers, additional carbon and vulcanizing agents such as sulphur to provide cross linking when the mix is cured. Thus, a long chain polymer is produced with widely varying physical properties according to its treatment, including a considerable ability to deform elastically, this characteristic being dependent upon the amount of cross linking. It is conventional to define characteristics by

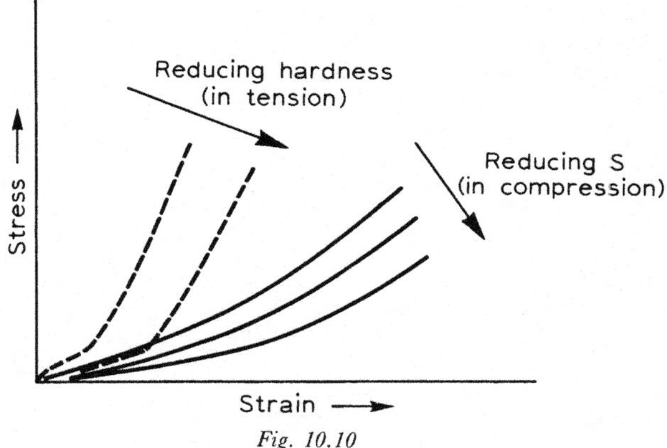

Fig. 10.10

hardness measurements, IRHD, °BS or SHORE A are typical scales, 0–100 and there is reasonable correlation with the elasticity theory at low strains, so that a number of formulae have been derived relating the elastic modulus to the penetration of an indentor. In terms of compression, performance is related to shape factor S, the ratio of loaded to force free areas. There is a suggestion of linearity up to 10% strain as shown in Fig. 10.10 and somewhat similar curves are typical for reducing values of hardness in tension and shear loadings.

A number of synthetic rubber compounds are available, Butadiene mixtures (SBR, NPR), Isobutylene, Polychloroprene (Neoprene) are typical. Some of these are cheaper than the modified natural rubbers discussed earlier but others are very expensive so that care must be exercised in measuring performance and economy.

EXERCISES

1. Distinguish between elastomers, fibres and plastics, and explain the mechanism of addition polymerisation and cross linking.
2. Compare the principal moulding processes for plastic materials and explain how inserts are incorporated.
3. Compare strength/weight ratios for the common plastic and ferrous materials and explain how the limitations imposed by fabricating methods and material properties are coped with in plastic constructions.
4. Discuss problems encountered in the machining of resins for photo-elastic specimens and the special manufacturing techniques which have to be used.
5. Define the important parameters of plastics in terms of high temperature applications and fatigue life.
6. How does the vulcanisation of rubber differ from the curing of plastic materials, what do you understand by mechanical hysteresis?

11

BEARING MATERIALS AND LUBRICANTS

In this chapter the basic principles of material and process selection will be extended to embrace problems of journal bearing design. It will be necessary to consider such design features in some detail since the choice of material may be dictated by the particular loading requirements and the availability of lubricant.

A simple choice chart in terms of life can be constructed similar to Fig. 11.1 to define the range of conventional bearing applications. The main consideration here will be in terms of the various types of plain journal bearings and bushings.

11.1. RUBBING BEARINGS

The simplest type of rubbing bearing involves the use of an un-lubricated bush, usually of non-metallic origin, for which a PV criterion [lbf/in² projected area × ft/min] is commonly quoted. This empirical product has little to commend it academically but it forms a simple design basis since it is easy to draw up a field of application, similar to Fig. 11.2 from which a particular working range can be extracted. It must be realised that a continuous wearing process is taking place so that the mating materials ought to have properties suitable for maintenance of a low wear rate. Considerable test data is available for particular wear rate criteria, 0·001 in/100 hr (0·00025 mm/hr) is a common parameter.

The simplest dry bearing materials are thermoplastics (often filled to improve properties or metal backed for strength) and woven fibre reinforced thermosets. More sophisticated types involve a fluorocarbon polymer overlay on an intermediate impregnated

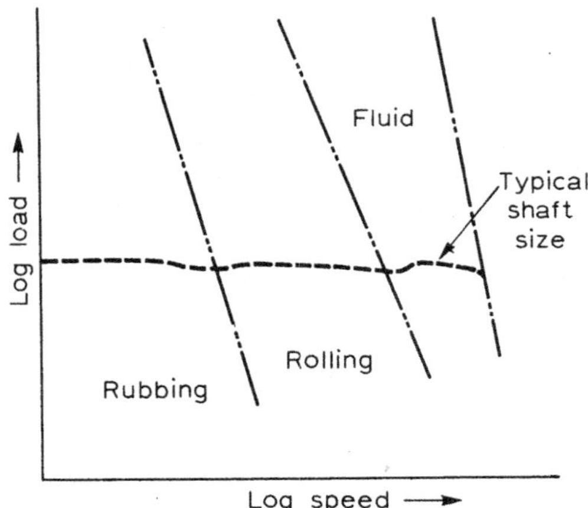

Fig. 11.1

bronze with a steel backing to provide adequate strength and some measure of heat dissipation. It is intended that a film of overlay will be transferred between the asperities of the mating surface, and alignment should receive particular consideration since there is no film of lubricant to spread the load. The bearings are usually supplied finish machined, with thin overlays it is not desirable to undertake further machining, so that shaft and housing must be prepared to suit correct running clearance and a suitable mounting fit respectively. Notice that plastic-faced materials require considerably larger running clearances, 0·005 in/in diameter is normal.

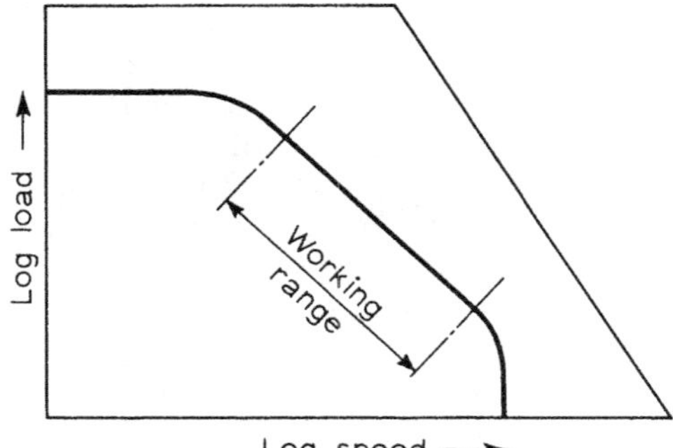

Fig. 11.2

11.2. POROUS BUSHES

There is a wide field of application for oil-impregnated porous metal bushing (essentially bronzes) intended to provide a self-lubricating effect. The sintering process, discussed in Chapter 4, is a vital phase in the production of these bearings after compaction under pressure of the alloying powders. These powder metallurgy processes are employed when the melting points of the alloys are so high that fusion techniques are impracticable, or when casting techniques could not

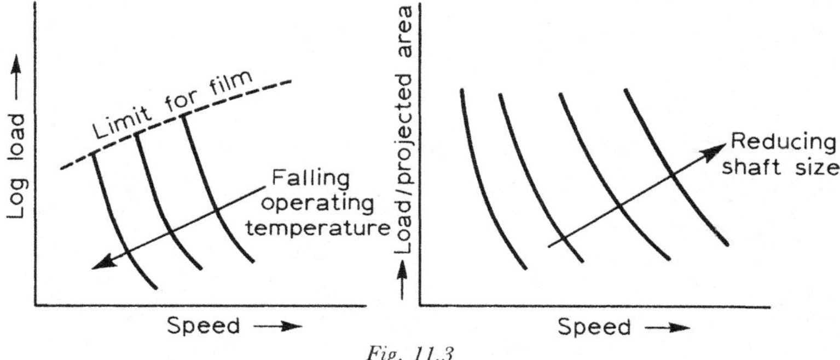

Fig. 11.3

provide the desired composition. The object is to provide compacts of porosity 25→35% in a powder mixture essentially 89% copper, 10% tin, 1% graphite.

Probably 90% of this porosity can be oil impregnated and, since retention is by capillary attraction, there must be a satisfactory relationship between pore size and surface tension. The mechanism of oil flow through these porous structures is complex and the real problem is whether the prediction of operating conditions has been accurate enough to remove any need for supplementary lubrication. Care must also be taken in mounting and supporting these thin-walled structures, which have less than half the strength of solid bearings of similar size, to allow for the different expansion co-efficients and the stresses imposed during fitting. As with the types reviewed in the previous section, there are restrictions on loading

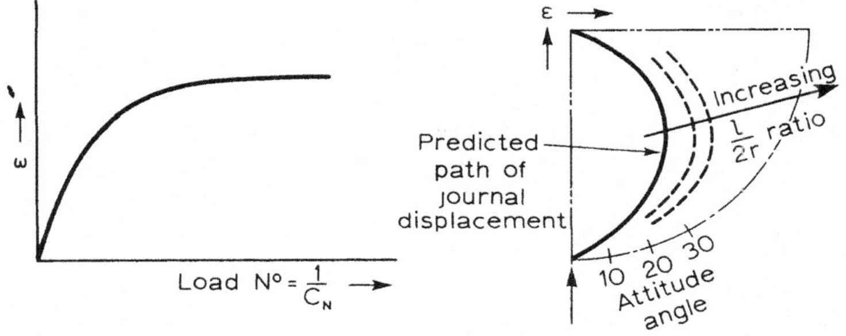

Fig. 11.4

due to problems of heat dissipation, although there is evidence of a pressure wedge and some oil circulation within the bearing if permeability is satisfactory.

The usual design intention is to operate at a much lower temperature than for full fluid lubrication, Figs. 11.3 and 11.4 show typical parameter relationships, while BS. 1131, Pt. 5 provides useful data on types and dimensions of liners and plain bushes commonly available.

11.3. TRIBOLOGICAL CONSIDERATIONS

Before considering the design of bearings in which hydrodynamic or film lubrication is the object (surface separated by a pressurised film due to journal rotation) it is appropriate to compare a real and an ideal bearing. This demands a review of the combined influence of friction, lubrication and wear, commonly thought of under the heading of tribology.

The ideal situation in a bearing envisages smooth and parallel surfaces, not subjected to thermal or elastic distortion, separated by a film of lubricant whose properties are at all times able to provide sufficient load carrying capacity to prevent metal to metal contact. In such circumstances material strength would be the only criterion, but in practice there is deflection, distortion and rubbing with penetration of the film by surface irregularities. The degree of failure in bearing performance is typified by generation of heat and size loss of mating parts, so for a real bearing it is necessary to consider complementary properties of the material pairs such as;

Compatibility, a measure of anti-scoring characteristics, a good facility here might mean acceptance of hardness to an unwanted degree.

Embeddability, to permit inclusions or debris being accepted by the softer bearing material.

Deformability, a rationalisation of hardness and elastic moduli to provide some compensation for misalignment and geometric error.

These factors are additional to the obvious necessity for the bearing material to resist corrosion attack from lubricant or atmosphere and any consideration of the effect of fatigue on bearing life.

11.4. SURFACE BEHAVIOUR

Surface interactions are determined by the real area of contact (often a small proportion of the apparent area) and friction is seen to occur for molecular as well as mechanical reasons. Behaviour of surfaces depends upon both volumetric and surface properties, the former embraces elastic and plastic parameters while surface performance is affected by the tendency to absorb from environment and the likelihood of acquiring a surface film. It is convenient to seek some relationship between elastic/plastic properties and test values. In terms of bearing metals parameters might relate yield strength to elastic modulus and penetration hardness. Although yield strength is a structure dependent property a reasonable correlation with test data is found by taking,

Yield strength $\simeq 0 \cdot 003E \simeq \frac{1}{3}$ penetration hardness in compatible units.

It is necessary to be discriminating in the use of such empirical relationships, values of the multiplying constants may be considerably affected by cold work prior to testing.

Since temperature at the formation of a junction due to intimate contact at speed is almost impossible to predict, it is difficult to be precise as to the surface energy of metals in a bearing or sliding situation. In solids, as with liquids, a surface situation arises where atoms have energy in excess of that found in the bulk of the material. The simplified relationship of solids,

$$\text{Surface energy} \propto (\text{penetration hardness})^{0 \cdot 33}$$

is often assumed. In conventional symbols $\gamma \propto H^{0 \cdot 33}$ and low values of γ/H are associated with better interaction behaviour, typically less adhesion, smaller wear particles and lower friction. In any event the majority of metals will form a surface film quite apart from any adsorbing activity.

Classical texts on wear of surfaces contain elaborate discussions as to the many explanatory theories which have been expounded. It will suffice to say here that *adhesive* and *abrasive* wear are the most common and there is a profusion of data on methods of reducing all but adhesive wear.

Adhesive wear is almost universal but rarely dangerous since the wear rate is low. Generalised laws for uncertain lubrication suggest a proportionality with applied load and distance over which sliding takes place, with an inverse relationship in terms of surface hardness. The lubrication available will inevitably condition the choice of material for bearing surfaces. Even when fluid lubrication is predictable under running conditions, the boundary situation will arise during starting and stopping so that the designer will choose materials to cope with a temporary departure from the ideal state.

Thus, the selection of one hard and one soft material typifies the choice for a journal bearing, with the intention of reducing the likelihood of seizure by local melting and the hope that wear debris will be implanted so as to eliminate further damage by circulation.

11.5. BEARING MATERIALS

The desirable hardness ratio of a metallic bearing pair might range from 5–14 over the range of relative soft bushings such as

babbitts or white metals→   copper/lead   →   lead bronze
(tin, antimony, copper)   (typically 70/30)   (75   / 20 / 5)
(copper/lead/tin)

These values are a useful guide in specifying a suitable journal hardness. Steel is the almost universal journal material and a surface finish $0.254 \rightarrow 0.381$ μm is suitable for the materials listed above, although special textures are necessary for some non-metallic pairings.

At the weaker end of the bearing range the facing material needs a stronger back, but the general use of stronger materials usually means a loss of desirable bearing attributes, since they invariably produce more heat in running and provide an additional function for the lubricant.

The white metal problem of considerable strength reduction at elevated temperatures is less significant in modern bearings where it is used mainly as a thin lining, and for low loadings it finds preference to copper/lead since the latter is less tolerant of debris and tends to increase journal wear. In modern practice, the use of gunmetals and phosphor bronze for bearings has been drastically reduced, designers prefer to adjust clearance, improve heat dissipation and increase oil flow so that softer faced bearings can be used whenever possible. This is particularly important if the boundary state is likely to be prolonged, damage might then be confined to one surface and the low melting point facility assists in alleviating local high pressure situations.

A useful table can be constructed for wear, seizure and corrosion ratings. If tin-base white metal is considered to have unity values for each of these a typical lead-bronze would be four times more susceptible to seizure and prone to attack, with the likelihood of causing four times more shaft wear (for a similar thickness) although the intrinsic strength of lead-bronze may be two and a half times greater. BS. 1131, Pts. 1–4, specifies ranges of plain and flanged metal bushings as wrapped bushes (from strip) and paired half liners. These are broadly based on copper alloys (BS. 1400) available either for machining in situ or pre-finished. The desired interference fit on outside diameter and circularity of bore is obtained when bushes are pressed into the housing.

11.6. VARIOUS DESIGN METHODS

Before undertaking the design of new bearings a careful review of the literature is desirable. Unfortunately the profusion of academic reports, many of which are not in terms of readily applicable data tends to confuse the young engineer. Referring to the book 'Engineering Design Problems', the solution to Project 2 involves an empirical estimation of the friction coefficient and bearing characteristics followed by equation of the heat terms to verify viscosity requirements, and determination of friction loss from charts prepared in terms of eccentricity ratio. As an alternative to that method the work of recent investigators based on the Sommerfeld solution of Reynolds equation will be briefly reviewed here.

The original postulation envisaged the creation of a hydrodynamic situation when a converging wedge-shape film is produced by motion of the surface relative to the lubricant and the simplified result of Sommerfeld was in terms of an infinitely long bearing so that end leakage could be neglected. In basic terms rf/c is equal to some function of

$$(r/c)^2 \; [un/p] = \text{Sommerfeld number } S, \text{ where}$$

r is journal radius, c is radial clearance, f coefficient friction, u absolute viscosity, n rev/sec, N rev/min, l length of bearing, p pressure on projected area, e eccentricity,

$$\text{eccentricity ratio } \epsilon = e/c = 1 - h_0/c$$

where h_0 is minimum film thickness and friction coefficient for a $360°$ bearing with clearance space filled with oil is conveniently written

$$f = (c/r) \; [(1 + 2\epsilon^2)/3\epsilon]$$

The *Ocvirk* solution for a short bearing makes the assumption of constant viscosity over a bearing arc of $180°$ and uses the term 'capacity number' C_N which is essentially

$$S(l/2r)^2$$

The results of experimental data are used to produce charts similar to Fig. 11.4 and show fair correlation with Sommerfeld results up to

$$l/2r \text{ ratios of 2, when } (2r/l)^2$$

term in load number is taken as unity for $l/2r$ values $1 \rightarrow 2$.

The work of *Raimondi* and *Boyd* followed a somewhat similar pattern but over a wide range of effective bearing arcs for centrally loaded bearings with $l/2r$ values of unity. Their proposals attempt to make some allowance for viscosity variation by using a mean bearing temperature and are usually presented as a series of step by step design charts.

A diagrammatic representation of typical performance curves is shown in Fig. 11.5 and their combination for a particular bearing might be presented as in Fig. 11.6. Notice how optimum clearance varies with load and that, at low loads, the maximum pressure is little affected by clearance. At high loads, peak pressures are concentrated on a smaller portion of bearing arc.

Any definition of an efficient design procedure should emphasize

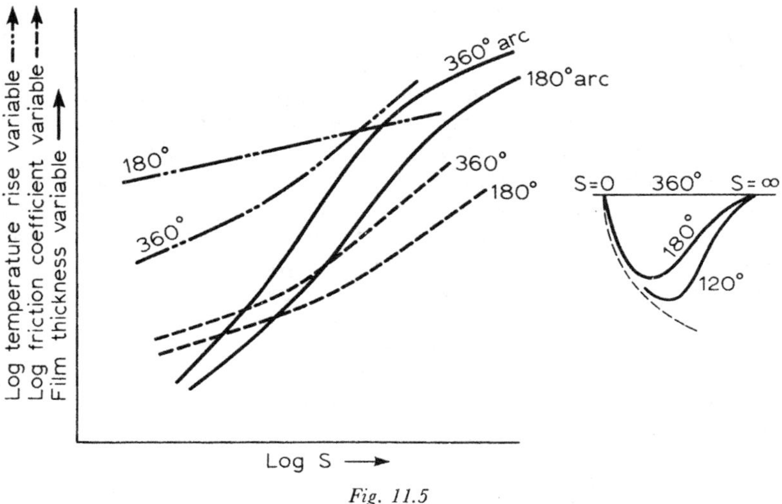

Fig. 11.5

the importance of adequate heat dissipation. Shortcomings in the prediction of bearing performance from empirical data are often due to insufficient consideration of the optimum relationships between material properties, lubricant behaviour and the maximum operating temperature which can be tolerated. For a whole bearing the assumption of average viscosity or that bearing temperature rise is half that of the lubricant gives reasonable correlation with test values of load carrying capacity for conventionally lubricated bearings.

11.7. ANALYSIS OF THERMAL EFFECTS

Analysis of thermal effects in the form of a balance, where heat terms are related to eccentricity ratio, can be used to derive a chart such as Fig. 11.7, from which a clearance ratio can be derived for

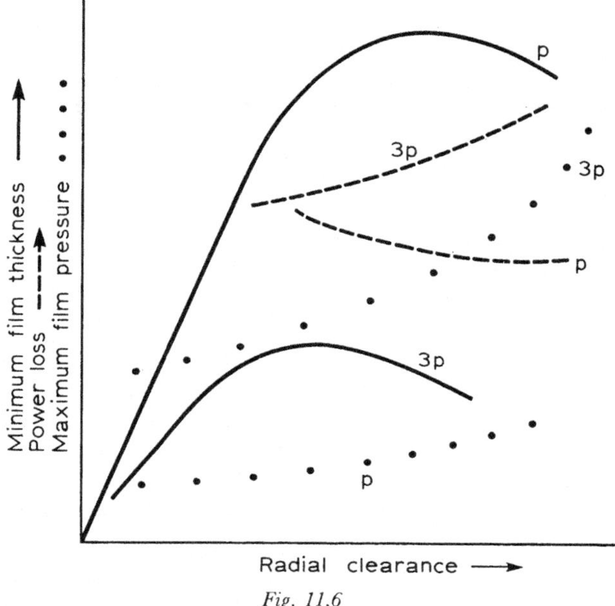

Fig. 11.6

an acceptable temperature difference. Reduction of friction for a given load is regarded by some designers as the most important criterion of performance improvement, the method adopted is to first consider reduction of clearance so that width of bearing can be reduced.

This method must allow, of course, for the minimum film thickness which can be tolerated and assumes satisfactory performance from the lighter oils which will have to be employed. Fig. 11.8 shows a typical chart of this type; notice a certain linearity at low loads and that for a given load the bearing stiffness increases as clearance is reduced.

A number of assumptions are contained in this tentative analysis but the preliminary results can often provide the starting point for a rigorous investigation when viscosity variation and extent of load carrying arc, etc., can be taken into account. A number of empirical values are quoted in the literature for acceptable clearances,

0·001 in/in dia is typical, but a much more important parameter is the minimum film thickness which must take into account any deficiency in surface texture of the mating pair. On the basis of a reamed bush and a ground shaft, say 2 in (50·8 mm) nominal size, the total irregularity of both surfaces might be 60 μ in (1·52 μm).

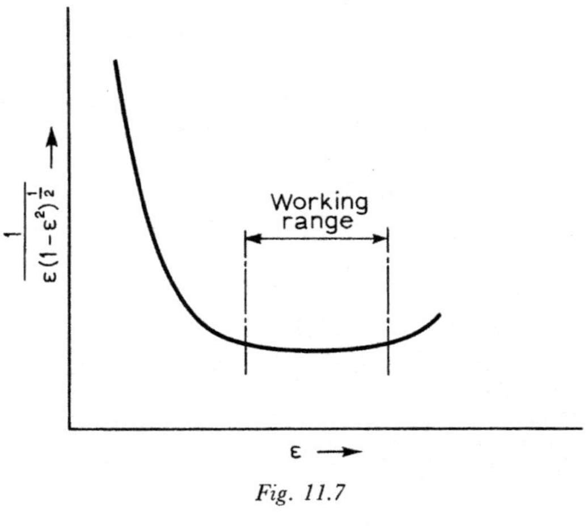

Fig. 11.7

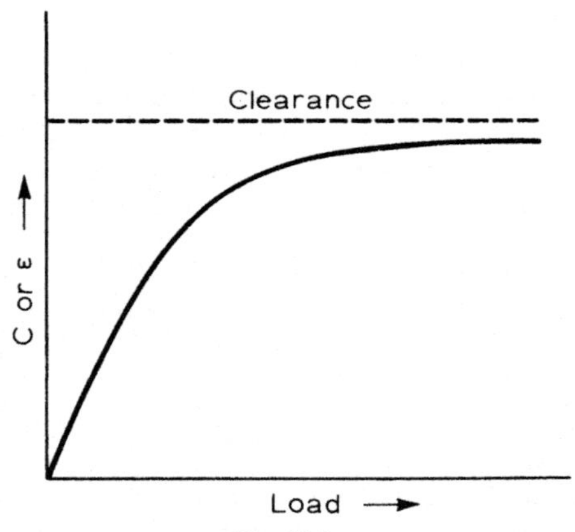

Fig. 11.8

A safe film thickness would be five times this, say 0·0003 in, and for a typical eccentricity ratio of 0·7 the diametral clearance from

$$\epsilon = 1 - 2h_0/C_D$$

gives $0·7 = 1 - 0·0006/C_D$ hence $C_D = 0·002$ in (0·05 mm). For conventionally prepared shaft and bush, without force fed lubrication, the film thickness should not be less than 0·00015 in/in dia.

11.8. DESIGN CHART METHOD

The design chart method is extremely popular and a recent Engineering Science Data publication 66023, recognising the more frequent use of pressure fed bearings, proposes solution methods for full bearings where a hydrodynamic situation may be expected to develop. Useful data on the effect of axial and circumferential grooving on performance is included for both Imperial and SI units. Some reserve should be exercised in the application of test data, much of this is derived from experimental work on machines whose rigidity would be quite impractical in commercial bearing applications. Such results sometimes predict extremely low friction co-efficients, based on accuracy and surface finish outside normal manufacturing specifications.

There has been considerable documentation on the correctness of the assumption to neglect the effect of pressure on component shape and viscosity of the lubricant. When contact pressures exceed $2\,316$ MN/m^2 the concepts of elastohydrodynamic lubrication need to be taken into account, point and line contacts in rolling elements and gears are typical examples, but the situation is unlikely to arise in conventional journal bearings.

11.9. LUBRICATION ARRANGEMENTS

In terms of lubrication techniques for metallic bearings, oil would commonly be regarded as first choice, although provided that the greasing regime is regular enough the reduced end leakage makes some self-lubricating systems popular. There is, however, the problem of variation in grease consistency with temperature and the much smaller facility for disposing of heat generated. The boundary situation must also be allowed for, a stationary journal might be thought of as resting on the microscopic high spots of a bearing surface. As motion starts or finishes there must be a period when there is no separation due to film; the surfaces which are wetted to a molecular level only slide over each other with a great risk of the minimal coating being penetrated by asperities. In such circumstances viscosity plays little or no part in frictional behaviour.

Location of feed is usually where pressure is least (unless force-fed) and, except where necessary for heavy loads at low speeds, it is desirable to avoid grooving in the lower half of the bearing because of its proximity to the maximum pressure area. Elaborate cross groovings are rarely desirable, they impede the formation of an oil film and provide an easy exit for lubricant from the pressure area.

In a one-piece radial bearing, a single longitudinal groove is often sufficient whilst, in a split bearing, a longitudinal chamber by bevelling the edges of cap and base will serve.

Assuming that sufficient area to carry the load and proper alignment has been provided, the most common type of bearing failure is due to inadequate clearance and the choice of an oil having insufficient body to meet the range of operating conditions. Temperature extremes due to poor ventilation at one end of the range or a cold situation where fluid is approaching its pour point, indicate the widely varying demands made upon a bearing lubricant. Oxidation of lubricants is always a problem, insoluble products may be deposited as sediment, clogging the oilways and retarding cooling due to their insulating effect. Lubricants which fulfill dual functions, such as in IC engines, often carrying abrading materials such as cylinder debris to bearing locations. Damage and the possibility of chemical attack on the bearings poses special material and lubricating problems.

The inherent usefulness of hydrocarbon oils as lubricants has been recognised for many years. They exhibit marked changes of viscosity for small temperature increments, but there has always been difficulty in classification, particularly in view of the wide range of additives now available. BS. 4231 classifies eighteen industrial oils; this is an extension of an original rationalisation to a series of twelve thought to cover the majority of bearing situations. Iterative calculations for bearing parameters obviously need a scheme which classifies in terms of a tolerance on viscosity at a fixed temperature rather than a range varying with temperature. Thus, the grade designations of BS. 4231 conveniently bear a close relation to kinematic viscosity in centi-stokes at 38°C, this temperature being chosen because it is one of the points on the Viscosity-Index System. Fig. 11.9 shows a temperature classification

	A	B	C
Viscosity C. St 100	32	68	150
A. S. T. M. (Saybolt)	150	315	700

Fig. 11.9

of plain mineral oils A, B, C, which are in common use.

A brief reference to suppliers lists suggests at least twenty manufacturers whose proprietary products will cover this range. The standard lists comparisons with S.A.E. data which have found wide use in this country, it is based entirely on viscosity range between defining temperatures of 0°F and 210°F.

EXERCISES

1. Discuss the use of a P.V. criteria for journal bearings manufactured from impregnated porous metals and indicate why it may not be a suitable basis when hydrodynamic lubrication is the object.
2. Discuss factors which control the choice of materials for bearing surfaces and list common materials in order of strength, wear and corrosion ratings.
3. Tabulate the common causes of bearing failure and explain how conventional performance is affected by alteration in lubricant properties and heat dissipation characteristics.
4. List the main causes of adhesive and abrasive wear and discuss the effect of lubricant behaviour in mitigation of surface damage due to wear phenomena.
5. Review various methods of surface protection against wear and corrosion with particular reference to the treatment of rubbing surfaces.
6. In terms of the discussion on plastic materials in Chapter 10, survey the field of application for dry bearing materials and list desirable characteristics for mating shafts.

12

BASIS OF SPUR TOOTH LOADING SYSTEMS AND MATERIAL CRITERIA

The topic of gear design is of special interest to the student and young engineer. Many of the mechanisms with which they have to deal include gear trains for which a selection of materials, an economic processing method and an appropriate pitch is required.

The simple method proposed in British Standards was amplified in several projects discussed in the book 'Engineering Design Problems' where systems involving spur, helical, bevel and worm gears were reviewed. Study of the vast literature on gear design however frequently baffles the young engineer. Many experienced investigators seem to decry the simple methods and propose elaborate modifications which are not always justified by clear advantages or are necessarily relevant to basic design requirements. This chapter will concentrate on spur gears and it is hoped to make a positive contribution by putting the various viewpoints into perspective. We will also show that a fundamental programme provides a satisfactory starting point from which a detail design plan, allowing for particular parameters such as corrected teeth and dynamic increments, can be developed.

12.1. LOADING CHARACTERISTICS

It will be necessary to consider events in different order from that used in earlier chapters where material and process selection went hand in hand. Most of the parameters in gear design are influenced by tooth profile; in particular, load carrying characteristics are identified by the assumption as to where the load acts, the number of pairs in mesh and how allowance for shape variation with different sizes can be made. Thus, we must first consider the geometry of teeth and how such proportions are relative to various terms in power formulae. The first lesson to be learned in consideration of real tooth contacts is that the concept of load coming on to a tooth at the theoretical pressure angle must be abandoned because of maldistribution of load. This problem was recognised in BS. 436 for so called 'commercial' gears and is best defined in terms of a load region which also makes allowance for addendum correction to be included as required, typically to cover the case of

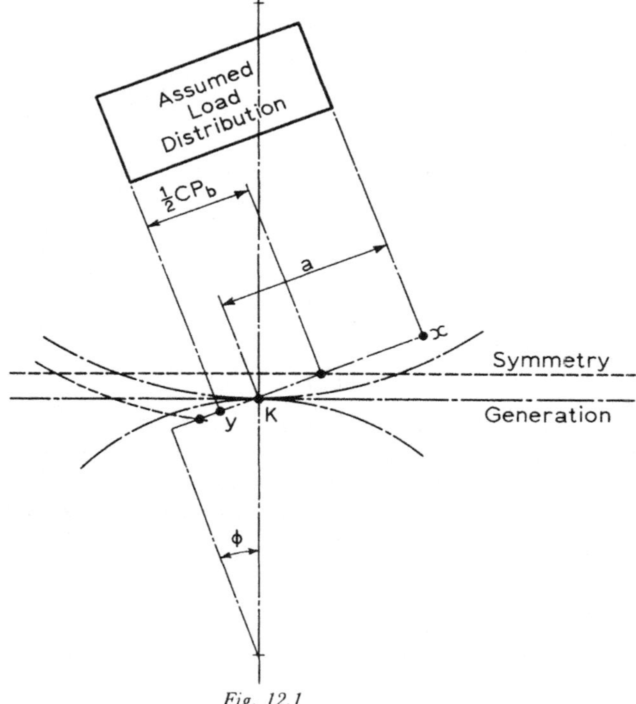

Fig. 12.1

non standard centres or to prevent undercutting, by varying the pitch line of the basic rack or cutter.

Fig. 12.1 shows the layout when the pitch line of generation is not the line of symmetry. Intersection of the latter with common tangent is used to define the extreme points of involute contact *xy*. These are not the theoretical positions defined by addendum circles but are conventionally taken as half the base pitch from coincidence point, i.e. the assumption is made that in the worst case the load is carried by one pair of teeth, in which any errors are ignored.

It is reasonable to regard the rectangular load diagram as defining the critical situation in terms of strength and wear. The variation of contact position for normal tooth load can be found with reference to Fig. 12.2 which is based on unit pitch.

For heavily loaded gears, or when extreme accuracy is desired,

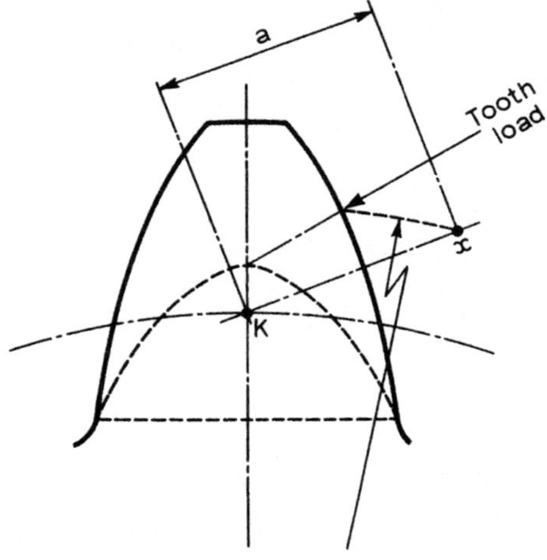

Fig. 12.2

the tooth profile modification system needs to take both deflection and pitch error into account. The load diagram now has the addition of triangular ends (Fig. 12.3) with the load points one base pitch from the effective ends of the common tangent thus defining the start of any profile modification.

12.2. POWER CALCULATIONS

The basis of a rating formula for allowable tangential load/inch of face is usually expressed as the product of a strength term S and a strength factor Y divided by diametral pitch P. Strength term is a function of basic material bending stress multiplied by a rotational speed factor X, the functional stress value is related to the ultimate tensile properties but is not a physical material test property, only a comparative value. Speed factor multipliers represent the relationship between permissible loading margins assuming that accuracy of particular gears is appropriate for their working speed range. Strength factors based on geometry and point of application are determined from $\dfrac{\text{tangential load}}{\text{resultant stress}}$, the stress pattern being based on the modified layout shown in Fig. 12.3 and since

$$\text{Torque } T = \frac{33000\text{HP}}{2\pi N} \quad \text{and} \quad \frac{\text{torque}}{\text{pitch radius} \times \text{face}} = \frac{\text{Allowable tangential}}{\text{load/inch width}} = \frac{SXY}{P}$$

then

$$\text{HP for strength} = \frac{SXYN \times \text{face} \times \text{No. teeth}}{126000\,P^2}$$

The problem of surface failure has proved more intractable than that of the bending situation, largely due to insufficient data on the fatigue strength of gear teeth and there is difficulty in correlating stress calculations with the endurance properties of various materials. The BS formulae for allowable tangential load/inch face follows the

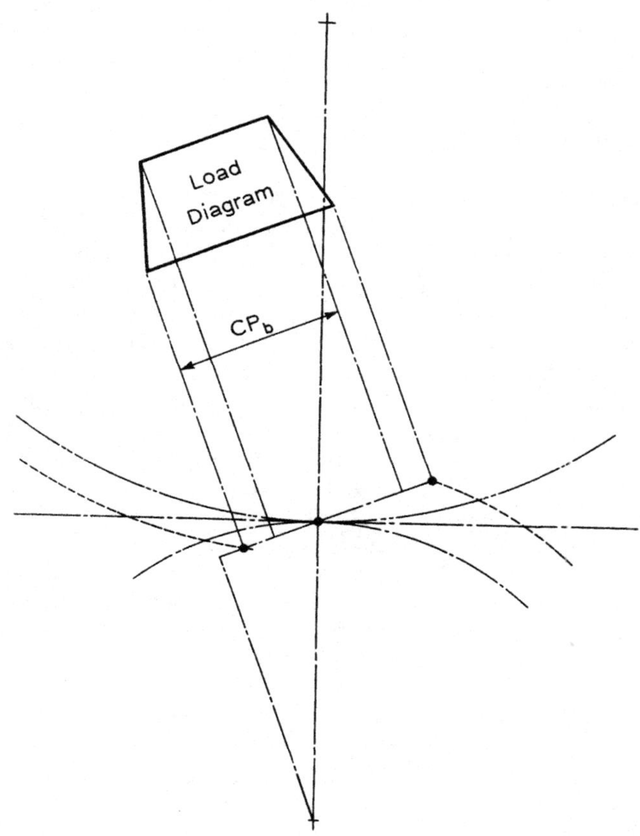

Fig. 12.3

same pattern as before, SXZ/K where S must involve some function of direct compressive stress, which it is tempting to relate to the proposals of Hertz for elastic materials. In many tables S is not a real stress value but merely a comparison between materials. However, instead of taking load as proportional to diameter, experience suggests the empirical expression of

$$S = \frac{\text{Normal tooth load/unit face}}{(\text{Relative radius curvature})^{0.8}}$$

relative radius $R = R_1 . R_2/(R_1 + R_2)$ and the criterion of relative load carrying capacity of different tooth number combinations of same pitch is expressed as the zone factor $Z = R.$ cos (pressure angle of engagement) for unit pitch and face. The shape of the load diagrams already referred to provide verification for the value of Z which depends on geometry of teeth and contact plane. In calculations the effect of different pitches is allowed for as the pitch factor $K = (P)^{0.8}$.

12.3. PARAMETER RELATIONSHIPS

It is necessary to examine in greater detail the apparently irrational nature of certain proposals which have been discussed. Since various parameters are related it is difficult to choose a straightforward pattern of assessment. The speed factor in bending purports to convert a basic material property, although not necessarily a function of repeated static load application on a section of changing shape, to a working stress. This factor ought to arrange for tooth loading to reduce with speed increase unless accurate profiles and meshing can be guaranteed.

Speed factors for various running times on logarithmic scales follow the pattern of Fig. 12.4. In terms of the common dynamic increment for geometrically similar gear pairs of different size moving at same pitch speed there is a sound argument for reverting to original concept and relating X values to velocity, although this will not directly make allowance for additional load repetitions on pinion. Errors in true profile meshing are caused by inaccurate shape, misalignment or elastic deformation. Loaded teeth are deflected in direction of motion and unloaded teeth are slightly behind their true position in relation to the others; tip thinning amount must exceed deflection and spacing error. The overall effect of these variations is to change relative velocities and cause a two peak load cycle with acceleration load being followed by an impact type reaction or dynamic load.

Precise results are difficult to define, since acceleration load is not a constant for elastic materials and the deformation variable depends on the number of pairs sharing the load. When the velocity is high and pitch is small the probable error is not reached and, for a given pair, there is a speed at which dynamic load is independent of the actual error. Thus, the commonly assumed proportionality

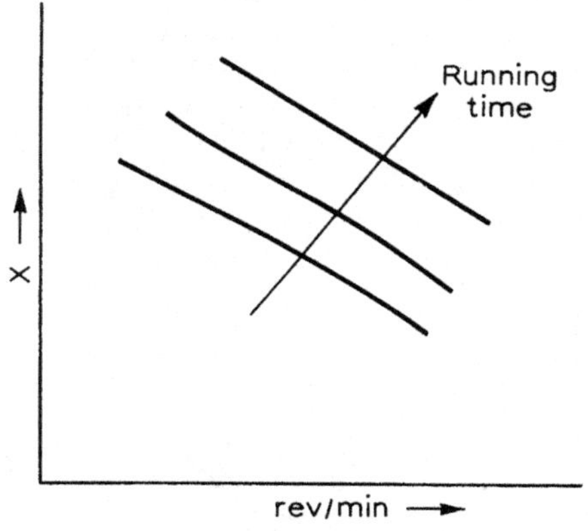

Fig. 12.4

between dynamic increment and load is in question and there seems a case for expression of a safe load after subtraction of the dynamic effect. Perhaps the best way of expressing a speed effect is by a relationship with elastic moduli; it could have similar values for hard and soft steels although their load capacity difference is large, the factor would be based on total number of load applications to cause failure.

The strength factor is based on precision gears and resultant compressive stress, for the doubtful assumption of tooth as a uniform cantilever. However the majority of failures originate on the tensile side of the tooth and there is scope for reshaping of factors on this basis, which would remove the dependence on number of teeth in mating gear.

Finally for the strength concept there is the question of stress concentration at the fillet. Many investigators have proposed multipliers for a range of probable fillet radii in terms of base thickness but these rarely allow for notch sensitive materials and the variation in individual behaviour of similar shapes produced by widely different processes. The proposals for wear capability are dominated by the surface stress term and, in particular, as to whether the equations of line contact based on isotropy and stress/strain proportionality can be applied to the dynamic situation.

Contact of this sort needs relating to the material of the mating gear and there are real differences of surface stress capacity between materials having the same tensile strength, thus the compressive stress at surface failure is unlikely to have a simple relationship with basic physical properties.

Material factors must acknowledge differences in fatigue behaviour so that endurance performance becomes a prime consideration. This is well documented for conventional machine components on a strength basis, but less is known for behaviour of pairs under high contact stresses, in particular it is difficult to postulate a fixed relationship between endurance limit and hardness. Tensile behaviour provides one key to such relationships as in Fig. 12.5

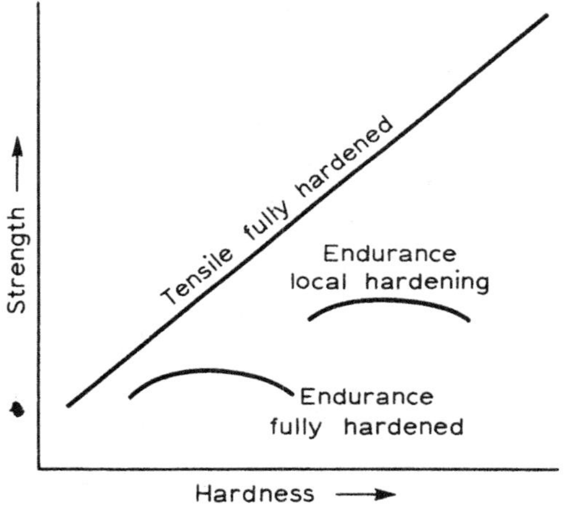

Fig. 12.5

which compares tensile and endurance performance; notice the effect of local hardening. Alternatively test data based on the number of load repetitions can be related to core hardness as in Fig. 12.6 and from this the basic surface stress factor can be derived. Speed factors for wear are more important than those for strength since the wear situation frequently governs overall dimensions.

The general shape of speed factor curves approaches $N^{-0.2}$ and from previous relationships,

$$\text{total tangential load } F \propto S.Z. \text{ face}/P^{0.8} . N^{-0.2}$$
$$\text{but } N \propto \text{Velocity} \times \text{Pitch/No. teeth } T$$
$$\text{so } F \propto S.Z. \text{ face. } T^{0.2}/PV^{0.2}$$

i.e. it is proportional to $V^{-0.2}$ and inversely proportional to P. This suggests dependence of wear speed factor on pitch line velocity, in any event the proportionality with a power of stress indicates the difference which should be reflected in any comparison with the strength speed factor. This dependence on $V^{-0.2}$ can be utilised to express a simplified zone factor, analagous to strength factor, which will not involve an empirical power of the relative radius of curvature, a term then introduced will indicate arbitrary rating differences between different gears.

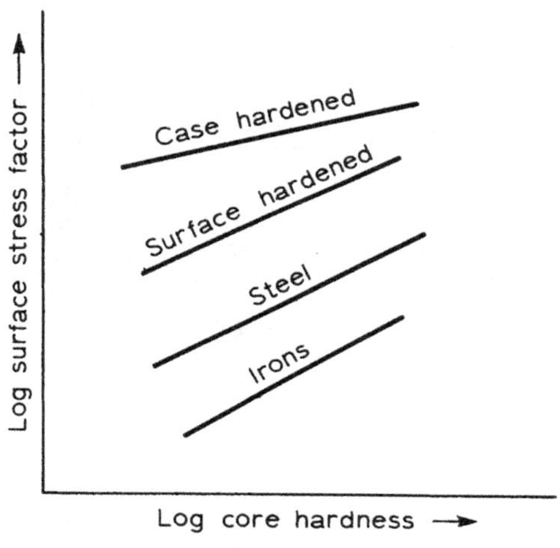

Fig. 12.6

12.4. ALTERNATIVE METHODS

It may be of interest to compare American rating methods with those already discussed and, as before, separate consideration of strength and wear is made. The basis of beam strength has its origin in the proposals of Lewis where the tooth was considered as a beam fixed at one end and loaded at the other and the bending load W depends on the form factor y derived from Fig. 12.7.

$$x/\tfrac{1}{2}b = \tfrac{1}{2}b/h \text{ so } x = b^2/4h$$

For face F and stress S the bending load $W =$

$$SFb^2/6h = (SF)2x/3 = (SpF)2x/3p = SpFy$$

where p is circular pitch so form factor $y = 2x/3p$

It is generally accepted that concept of a tip load is not realistic and considerable documentation is available on the effect of deflexion and manufacturing error for which Fig. 12.8 are typical. The empirical equation for dynamic load is

$$0.05V(bC + F_T)/[0.05V + \sqrt{(bC + F_T)}]$$

where V is pitch line velocity

C depends on shape, action error and material and the value of dynamic load·is added to the tangential load F_T to find the basic strength requirement.

In terms of a limiting load for wear the Hertz theory of contact between two cylinders is a useful basis, except that radius of curv-

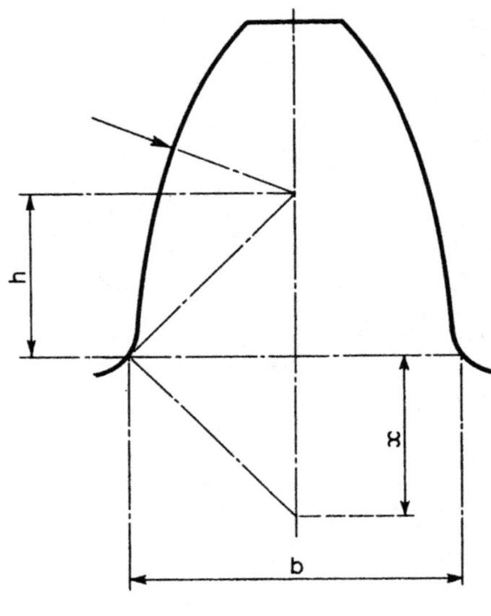

Fig. 12.7

ature is constantly changing so that a specific position must be designated. Wear is often first apparent at pitch line and the basic equation Endurance stress

$$S^2 = 0.35 \times Wear\ Load\ [1/r_1 + 1/r_2]/(Face\ F)^2\ [1/E_1 + 1/E_2]$$

can be rewritten as

$$Wear\ Load = D_1.F.Q.[S^2\ Sin\ \varnothing\ (1/E_1 + 1/E_2)/1.4]$$

where Q is tooth number ratio $2N_2/(N_1 + N_2)$

(suffix $_1$ pinion; $_2$ wheel) and square bracket [ ] is load stress factor K Thus we may write Wear Load $= D_1 FQK$ and the usual design method ensures that this value will exceed the strength rating. This concept places great importance on the allowable stress, conventionally based on the endurance limit for released loading ($\frac{1}{3}$ UTS is typical) but modified by a speed factor. An empirical form for surface endurance limit is 400 (average Brinell hardness) $- 10\ 000$ lbf/in².

This section would be incomplete without some reference to the many methods which have been proposed to allow for non-uniform running. BS. 436 proposals are for a normal rating of 12 hrs/day for 26 000 hours although few industrial operating schedules involve prolonged running without some variation. The real criterion is the number of cycles at which an endurance limit is reached, 10^7 may sometimes be assumed for bending stresses but the limit for a surface

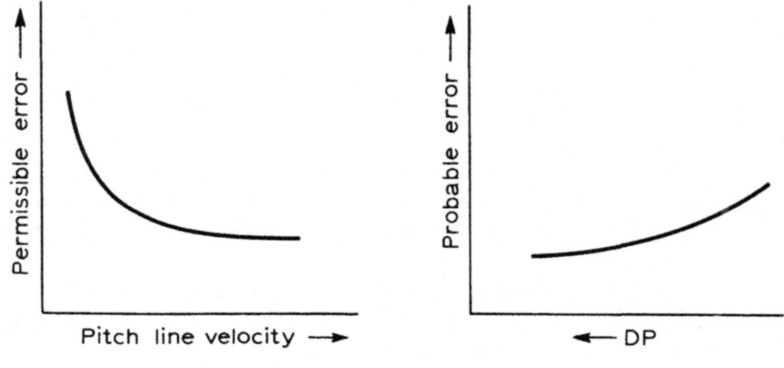

Fig. 12.8

stress condition is very much higher. For variable loading it is customary to determine an equivalent running time at uniform load which produces the same effect, special attention must be paid to momentary loads. Rules for variable cycles tend to be inconsistent and rarely produce a satisfactory basis from which a safety factor can be determined.

12.5. FAILURE CONCEPTS

The general lubrication situation has been discussed in the previous chapter, in fact a bearing analogy is sometimes used to typify lubrication between sliding surfaces as in gear teeth. Remember however that there is duality of motion, similar to a journal and bearing rotating at different speeds, and considerable sliding at the onset of mesh. Rolling and sliding aid in film formation, although since the wedge formed between contacting surfaces is much shorter than in a bearing, the fluid film is more difficult to sustain unless there is an ample supply of lubricant. Oil is in general the first choice for gear lubrication because of its better prospects of limiting the extent of boundary lubrication. Careful attention to the viscosity value in terms of sliding velocity and operating temperature will usually result in a mineral oil selection, with bath lubrication provided at lower speeds and spray or jet alternatives when turbulence might be a problem or additional cooling is to be provided.

In terms of material behaviour, tooth failures can be classified under the headings *pitting, abrasion, scoring, flaking* or *spalling, scuffing* or *galling*. In a *pitting* situation, variations in surface conditions have resulted in uneven load distribution and concentration of stress adjacent to the pitch line. Low rates of slide to roll are suspect and the tiny pits which are formed serve as the origin of surface fatigue phenomena.

Abrasion and *scoring* are to a large extent independent of the efficiency of the lubrication regime. The first of these is due to the entry of airborne particles, which emphasizes the importance of surface hardness, while scoring is accentuated by wear debris from local welding. In both cases the immediate effect is a breakdown in the fluid film.

Flaking and *spalling* are commonly accepted as a fatigue problem, usually found on flanks of teeth since load is greatest at start and finish of mesh, speed is the criterion.

Scuffing or *galling* failures are characterised by metal removal due to failure of oil film to carry the load. Starting at pitch line where film formation tendency is most critical the scuffed area is transferred above and below depending on rotation and leads eventually to a boundary situation. This, in turn, engenders conditions favourable to the other types of failure mentioned earlier.

Before considering gear materials we can now summarise the enquiry into rating methods which has occupied the first part of this chapter. It seems fair to say that, for commercial gears, the results obtained from the simple methods are always likely to provide a conservative result. Some designers, of course, mistrust a substitution system whose parameters are so dependent and in particular, that after some illogicality in analysis, the overall accuracy is placed further in question on multiplication by a 'stress' value of doubtful origin. In this type of design exercise the existence of a variable safety factor is just as worrying as the specification of unsafe dimensions, quite apart from the uneconomic aspects of over-designing.

For special designs the analysis of loading method provides a starting point from which particular modified profiles can be determined to provide the best possible running conditions. In any real design situation the first choice is of gear type in terms of load and speed which pre-disposes the number of pinion teeth for the materials available. The need for addendum modification should then be investigated, typical size ranges for blank or tip diameter will express the extremes between undercutting and pointed teeth, and will indicate the latitude in centre distance.

Design charts such as Fig. 12.9 are most useful to indicate the start of a geometric design procedure and the permissible torque for a given material combination can be defined by a chart such as Fig. 12.10, this would be based on unit diameter.

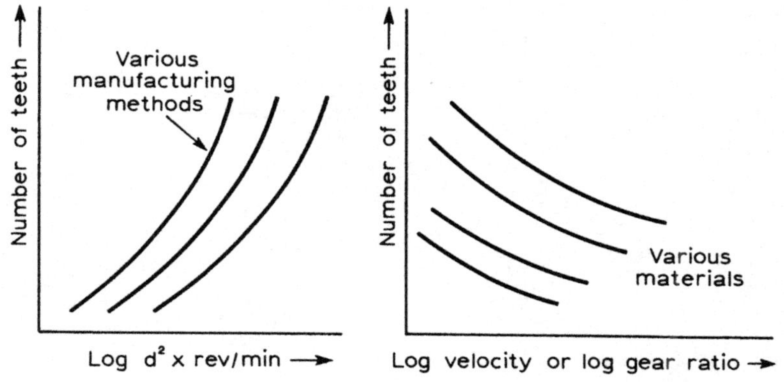

Fig. 12.9

Fig. 12.10

12.6. MATERIAL CONSIDERATIONS

The majority of gears are made from metal, and the choice is not always simple, depending as it does on composition, property amendments obtained by subsequent heat treatment and the manufacturing process available. Hardness exerts a significant influence, easy cutting is possible at 200 Brinell whereas at 500 grinding might be required to finish and to achieve higher hardness some type of surface process is necessary after machining. The relationship between hardness and carbon content is shown in Fig. 12.11,

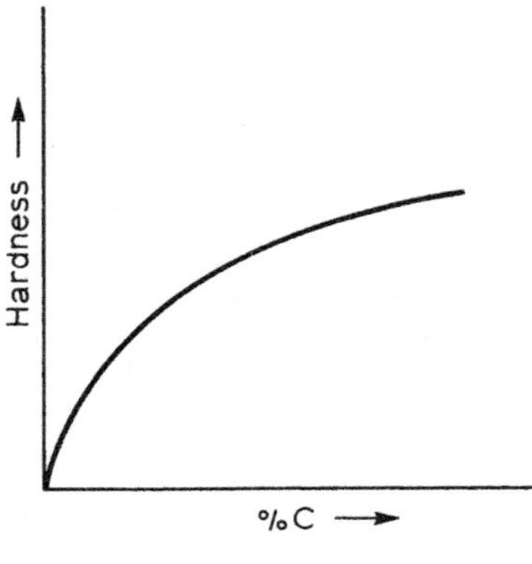

Fig. 12.11

approximately 6% carbon is necessary for a fully hardened material to reach 600 Brinell.

Adequate strength for a particular application may be provided at a low carbon content but wear resistance could be suspect because of deficiency in surface carbides. The increase in load carrying capacity from the use of high tensile through hardened materials is relatively small in the range of simple manufacturing processes, a Brinell range 300–350 frequently defines adequate strength and is also a commonly quoted limit for good machinability.

Cast iron specifications to BS. 1452 have been discussed earlier, grades 12 and 14 are commonly used for gears as are the three grades of BS. 821 providing ultimate tensile ranges between 154 → 340 MN/m². The cheapness of low- and medium-grade irons is attractive and they are prepared as hard as possible consistent with machinability, 170–220 Brinell is typical. Nodular and innoculated irons are widely used where compatible behaviour and size retention is important, but their considerably higher cost needs justification unless their improved tensile properties are fully utilised. Cast carbon steels need medium carbon content, 0·35–0·45% carbon for wear properties and careful heat treatment to refine grain structure. The use of alloyed materials for higher strengths may be precluded due to distortion of difficult shapes in quenching, an alternative is the preparation of a rolled rim and a fabricated construction.

The four basic types of gear steels listed in BS. 436 are the cast, forged, surface-hardened or case-hardened varieties with specifications as BS. 970 and 971. Notice that a given specification may be provided in alternative conditions, thus EN.8 as a proved plain carbon gear steel is available in a forged (normalised) version or surface hardened; in the former state maximum bending stress capacity is found, but its surface resistance is half that of the surface hardened version. Case-carburised or skin-hardened materials provide a good combination of a tough and ductile core with optimum surface conditions although distortion during heat treatment is sometimes a problem, EN.34 and 36 are typical choices. Nitriding is often used, teeth are exposed to ammonia gas whose surface penetration combines with alloys to form nitrides. The attraction is the retention of core properties which have already been treated to the desired condition. EN.40B provides the hardest skin; typical variation is shown in Fig. 12.12. The induction

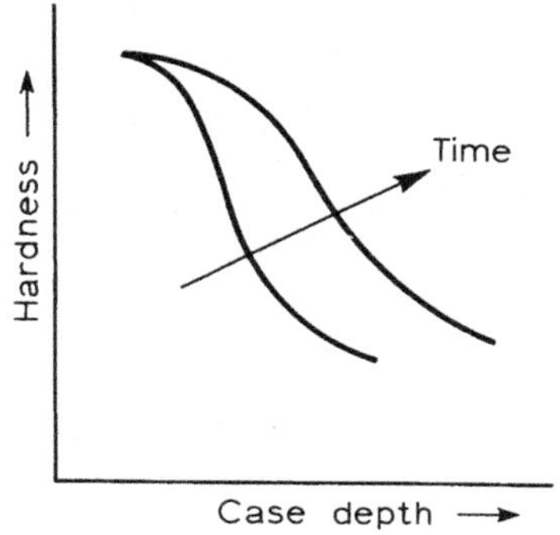

Fig. 12.12

hardening method is often applied to medium carbon gears of large pitch, it provides a through hardening type of effect on the flanks and root of teeth when jet quenched, EN.24 is typical.

Non-ferrous materials are not widely used as spur gears, mainly on the grounds of expense, although certain bronzes have unique qualifications for tooth loading and appear in other tooth forms. Light alloys generally need facing to improve wear life. A life based surface stress is a useful criterion and should consider material pairings for best results.

Fig. 12.13 shows how a typical range chart for geometric design might be established.

In the non-metallic range the field is dominated by thermoplastics such as nylon and phenolic-laminated thermo-setting resin bonded fabrics. These types are rarely critical in terms of surface stress but pose particular problems in terms of bending and wear. Allowable load/unit face can be calculated on a similar basis to metallic types but corrections must be introduced for temperature, humidity and

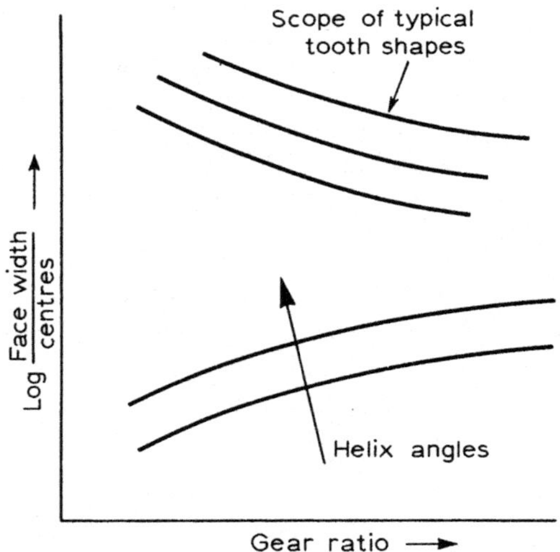

Fig. 12.13

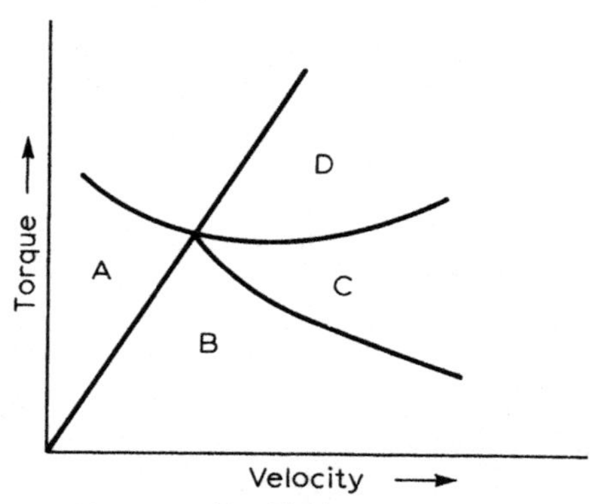

Fig. 12.14

wear. The low thermal conductivity of plastic materials poses load carrying problems unless suitable lubrication can be organised to carry away the heat generated. A chart such as Fig. 12.14 is useful to define areas such as A where it is difficult to form an oil film, or C where abrasion of the harder material may penetrate the film.

12.7. PRICE COMPARISON, OPTIMISATION

Assuming that ruling section restrictions have been satisfied with adequate strength and that an economic cutting process is available, the choice of material reflects the dual criteria of price and compatibility. A typical price index for the materials mentioned earlier, taking ordinary grade cast iron as unity will range; EN 8/9 1·25, Malleable Iron 2, Cast Steel 2·5, EN 27 4·5, EN 34 5, EN 40B 9, Phos. Bronze 12·5. Compatibility data is largely amassed from case studies, and it is generally agreed that pinion be harder than wheel to cover the larger number of load repetitions. A first choice for moderate loading might be EN 9 pinion, EN 8 wheel with EN 27/ EN 9 providing a higher strength alternative whose improved qualities are obtained at about the same percentage increase in cost. Alternately, a rating table can be prepared in terms of bending or surface stress capability, for the latter typical pairings have relative values as tabled in Fig. 12.15.

The selection of materials and processes in power transmission applications provides an excellent opportunity to use the optimum

Pinion	En number	8	9	9	9	27	34	40B
	Rating	1	1·5	1·5	1·5	2·2	7	5·3
Wheel	En or type	CI(12)	Phenolic resin	Nylon	En8	9	19	19
	Rating	·7	·5	·9	1	1·5	4·6	4·6

Fig. 12.15

design method. The relationship between gear geometry and manufacturing methods might be a typical example of such an investigation. Let us assume that an optimum value for power transmission capability is to be expressed for a reduction pair in terms of known values for pinion speed, velocity ratio and centres, with data available as to various manufacturing parameters.

A *primary design equation* will express horsepower in terms of tangential tooth load, size and velocity. Subsidiary equations will include allowance for errors and elasticity in terms of dynamic increments. Limiting equations might be applied for diameter (because of centres), for facewidth (contact length), for cutter pitch available, for helix angle (thrust restriction) and for permitted load to avoid overstressing in terms of strength or surface conditions. The primary equation must now be developed, to include subsidiaries and limits, into a form involving elastic moduli and either a strength or a form factor term. Thus the optimum materials for maximisation of strength and wear capacity will be seen by inspection, although there may be relevance between the factors since an expression defining form is a function of hardness which means some relationship with strength and thus a criterion of failure.

EXERCISES

1. Explain why hardness is an important parameter in the selection of gear materials and categorise the various methods of tooth hardening for spur gears.
2. Indicate the scope of the various profile finishing processes and compare their performance in terms of economic considerations.
3. Discuss how the assumption of a simple relationship between static load and dynamic effects has to be modified to allow for gear contacts.
4. Consider the various mechanisms of gear failure and explain how these are influenced by alternative lubrication regimes.
5. Compare the British and American tooth load rating methods for spur gears showing the basis of formulae derivation.
6. Given details of a simple gear pair in terms of size, hp and angular velocity, conduct an investigation into the optimum design of a gearbox casing, allowing for various shaft positions, on the basis of minimum weight.

BIBLIOGRAPHY

British Standards

BS.   15:1961.   Mild steel sections for structural purposes.
BS.   18:1962.   Methods for tensile testing of metals.
BS.   240:1962.   Method for Brinell hardness test.
BS.   308:1964.   Engineering drawing practice.
BS.   309:1958.   Whiteheart malleable iron castings.
BS.   310:1958.   Blackheart malleable iron castings.
BS.   427:1962.   Method for Vickers hardness test.
BS.   436:1967.   Pt. 1. Machine cut gears, helical and straight spur.
BS.   600:1960.   Application of statistical methods to industrial standardisation and quality control.
BS.   891:1962.   Method for Rockwell hardness test.
BS.   970:1955.   Wrought steels, bars, bullets and forgings; EN Series.
BS.   971:1950.   Commentary on BS wrought steels, EN Series.
BS. 1004:1955.   Zinc alloy die castings.
BS. 1131:1955.   Pts. 1–5. Plain bearings (metal).
BS. 1313:1947.   Fraction defective charts for quality control.
BS. 1400:1961.   Copper alloy ingots and copper and copper alloy castings.
BS. 1452:1961.   Grey iron castings.
BS. 1470:1963.   Wrought aluminium and aluminium alloys for general engineering purposes.
BS. 1490:1963.   Aluminium alloy ingots and castings for general engineering purposes.
BS. 1775:1964.   Steel tubes for mechanical, structural and general engineering purposes.
BS. 1916:1953.   Pt. 1–3. Limits and fits for engineering.
BS. 2026:1953.   Tolerances for moulding in thermosetting materials.
BS. 2094:1954.   Glossary of terms relating to iron and steel.
BS. 2564:1955.   Control chart techniques when manufacturing to a specification.
BS. 2590:1955.   Acceptance tests for sintered metal components.
BS. 2782:1965.   Methods of testing plastics.
BS. 2789:1961.   Spheroidal or nodular graphite iron castings.
BS. 2970:1959.   Magnesium alloy ingots and castings.
BS. 3100:1967.   Steel castings for general engineering purposes.
BS. 3146:1959.   Investment castings in metal.
BS. 3179:1957.   Comparison of British and overseas standards for steels.
BS. 3370:1961.   Wrought magnesium alloys for general engineering purposes, sheet and strip.
BS. 3373:1961.   Wrought magnesium alloys for general engineering purposes, bars and sections.
BS. 3500:1962.   Methods for creep and rupture testing of metals.
BS. 3518:1962.   Methods of fatigue testing.
BS. 3800:1964.   Methods for testing the accuracy of machine tools.

BS. 4042:1966.   Tolerances for injection mouldings in thermoplastic materials.
BS. 4114:1967.   Dimensional and quantity tolerances for steel drop and press forgings.
BS. 4231:1967.   Viscosity classification for industrial liquid lubricants.
BS. 4360:1968.   Weldable structural steels.

Publications of the Engineering Sciences Data Unit, London

ESD. 65007.   General guide to the choice of journal bearing type.
ESD. 66001.   Geometric design of spur and helical gears.
ESD. 66023.   Calculation methods for steadily loaded pressure-fed journal bearings.
ESD. 68001.   Spur and helical non-metallic gears, materials and load capacity.
ESD. 68018.   Guide to design and material selection for dry rubbing bearings.
ESD. 68040.   Spur and helical gears, choice of materials and estimate of dimension.

Books and Articles

4th International Machine Tool Conference, Manchester. *Design of Machine Tool Structural Elements.* Pergamon (1963).

Allen, P. W., Lindley, P. B., and Payne, A. R., *Use of Rubber in Engineering.* Maclaren (1967).

Ashford, F. C., *Designing for Industry.* Pitman (1955).

ASME Handbook. *Metals Design Engineering.* McGraw-Hill (1953).

Begeman, M. L., and Amstead, B. H., *Manufacturing Processes.* John Wiley (1963).

Billmeyer, F. W., *Textbook of Polymer Science.* Interscience, New York (1962).

Blodgett, O. W., *Design of Weldments.* Lincoln Welding Foundation, Cleveland, Ohio (1963).

Boothroyd, G., *Fundamentals of Metal Machining.* Arnold (1965).

Conway, H. G., *Engineering Tolerances.* Pitman (1962).

Copper and its Alloys. Copper Development Association (1960).

Cottrell, A. H., *Introduction to Metallurgy.* Arnold (1967).

Dubois, G. B., and Ocvirk, F. W., *Experimental investigation of Eccentricity Ratio, Friction and Oil Flow of Journal Bearings.* NASA. Technical Note 3491 (1955).

Dudley, A. W., (ed.), *Gear Handbook.* McGraw-Hill (1962).

Flinn, R. A., *Fundamentals of Metal Casting.* Addison Wesley (1963).

Freeman, G. G., *Silicones.* Iliffe (1962).

Gregory, E., and Simons, E. N., *Heat Treatment of Steels.* Pitman (1963).

Higgins, R. A., *Engineering Metallurgy (Pt. II).* E.U.P. (1960).

Hoffman, O., and Sachs, G., *Introduction to Theory of Plasticity for Engineers.* McGraw-Hill (1953).

Houghton, P. S., *Heat Treatment of Metals* (2 Vols). Machinery Publishing Co. (1960).

Institute of Metallurgists. *Heat Treatment of Metals*. Iliffe (1963).

Jevons, J. D., *Metallurgy of Deep Drawing and Pressing*. Chapman and Hall (1940).

Johnson, R. C., *Optimum Design of Mechanical Elements*. John Wiley (1961).

Jones, W. D., *Fundamentals of Powder Metallurgy*. Arnold (1960).

Keyser, C. A., *Materials Science in Engineering*. C. E. Merrill, Ohio (1968).

Kobayshi, A., *Machining of Plastics*. McGraw-Hill (1967).

Koenigsbeger, F., *Design Principles of Metal Cutting Machine Tools*. Pergamon (1964).

Larke, E. C., *Rolling of Strip Sheet and Plate*. Chapman and Hall (1963).

Lipson, C., *Wear Considerations in Design*. Prentice Hall (1967).

Lissaman, J., and Martin, S. J., *Principles of Engineering Production*. E.U.P. (1966).

MacMaster, R. C., (ed.), *Non-destructive Testing Handbook* (2 Vols). Ronald Press (1959).

Magnesium in General Engineering. Magnesium Industry Council.

Matousek, R., *Engineering Design—a Systematic Approach*. Blackie (1966).

Merritt, H. E., *Gears*. Pitman (1955).

Palin, G. R., *Plastics for Engineers*. Pergamon (1967).

Pascoe, K. J., *Introduction to Properties of Engineering Materials*. Blackie (1961).

Pearson, C. E., and Parkins, R. N., *Extrusion of Metals*. Chapman and Hall (1960).

Peck, H., *Allocating Tolerances and Limits*. Longmans. (1968).

Peterson, R. E., *Stress Concentration Design Factors*. Wiley (1966).

Practical Guide to the Design of Grey Iron Castings. Council of Ironfoundry Associations, London (1956).

Properties of Aluminium and its Alloys. Aluminium Development Association (1955).

Raimondi, A. A., and Boyd, J., *A Solution for the Finite Journal Bearing, Application to Analysis and Design*. ASLE Transactions, Vol I(i) (1958).

Ritchie, P. D., (ed.), *Physics of Plastics*. Iliffe (1965).

Rollason, E. C., *Metallurgy for Engineers* (3rd edn.). Arnold (1968).

Sachs, G., *Fundamentals of the Working of Metals*. Pergamon (1954).

Sharpe, H. J., (ed.), *Engineering Materials—Selection and Value Analysis*. Iliffe (1966).

Simonds, H. R., (ed.), *Encylopaedia of Plastics Equipment*. Reinhold (1964).

Skeist, I., (ed.), *Handbook of Adhesives*. Reinhold (1963).

Sommerfeld, A., *Hydrodynamic Theory of Lubrication* (Translation). Z. Math Physick, Vol 50, pp. 97–155 (1904).

The Design and Properties of Steel Castings. British Steel Castings Research Association, London (1962).

Thomas, L. F., *The Control of Quality*. Thames and Hudson (1965).

Tuplin, W. A., *Gear Design*. Machinery Publishing Co. (1962).

Von Alven, W. H., (ed.), *Reliability Engineering*. Prentice Hall (1964).

Wood, C. W., and Halliday, A. K., *Organic and Physical Chemistry* (2 Vols). Butterworth (1963).

INDEX

Printed in Dunstable, United Kingdom

85047107R00080